PRAISE FOR T

'Like Jared Diamond and Stephen Jay Gould, Tim Flannery has the ability to take complex ideas and—seemingly effortlessly—make them accessible.' *Sydney Morning Herald*

'For decades Flannery's essays and books have offered crucial summaries and clarifying perspectives on global biodiversity, climate, Australian history, mammals, megafauna and possibilities for alternative energy sources...*Life: Selected Writings* is a testimony to [his] independent brilliance, and also to his canny talent for combining the high pitch and the common touch.' *Australian*

'This man is a national treasure, and we should heed his every word.' *Sunday Telegraph*

'Tim Flannery is a man of dazzling talents: scientist, climate change activist, academic, explorer.' Noted NZ

'[Flannery] is a master storyteller, with an eye for the revealing detail.' *Australian Book Review*

'Tim Flannery is the real thing: a man with a gift for lucid exposition, who can really make his subject come alive.' *Literary Review* [UK]

'Flannery's writerly art stirs the imagination to pay attention.' *Australian Literary Review*

'Flannery [is] a writer who sneezes at political correctness and charges into the densely land-mined territory of the biological determinants of human behaviour.' *Washington Post*

'Flannery synthesizes a vast range of scientific studies and a decent selection of historical and cultural writings, leavening those with his own forceful ideas.' *New York Times Book Review*

'Flannery has a great ability to distil complex subject matter into something you can wrap your head around.' *North & South*

'No one tells it better than Tim Flannery.' David Suzuki

'If you are not already addicted to Tim Flannery's writing, discover him now.' Jared Diamond

ALSO BY TIM FLANNERY

Mammals of New Guinea

Tree Kangaroos: A Curious Natural History with R. Martin, P. Schouten and A. Szalay

The Future Eaters

Possums of the World: a Monograph of the Phalangeroidea with P. Schouten

Mammals of the South West Pacific and Moluccan Islands

Watkin Tench, *1788* (ed.)

John Nicol, *Life and Adventures 1776–1801* (ed.)

Throwim Way Leg: An Adventure

The Explorers (ed.)

The Birth of Sydney (ed.)

Terra Australis: Matthew Flinders' Great Adventures in the Circumnavigation of Australia (ed.)

The Eternal Frontier

A Gap in Nature with P. Schouten

John Morgan, *The Life and Adventures of William Buckley* (ed.)

The Birth of Melbourne (ed.)

Joshua Slocum, *Sailing Alone around the World* (ed.)

Astonishing Animals with P. Schouten

Country

The Weather Makers

We Are the Weather Makers

An Explorer's Notebook

Here on Earth

Among the Islands

The Mystery of the Venus Island Fetish

Atmosphere of Hope

Sunlight and Seaweed

Europe

Life

Tim Flannery is a scientist, an explorer, a conservationist and a leading writer on climate change. He has held various academic positions including visiting Professor in Evolutionary and Organismic Biology at Harvard University, Director of the South Australian Museum, Principal Research Scientist at the Australian Museum, Professorial Fellow at the Melbourne Sustainable Society Institute, University of Melbourne, and Panasonic Professor of Environmental Sustainability, Macquarie University. His books include the award-winning international bestseller *The Weather Makers*, *Here on Earth* and *Atmosphere of Hope*. Flannery was the 2007 Australian of the Year. He is currently chief councillor of the Climate Council.

The Climate Cure

Solving the Climate Emergency in the Era of COVID-19

Tim Flannery

Text Publishing Melbourne Australia

textpublishing.com.au

The Text Publishing Company
Swann House
22 William Street
Melbourne Victoria 3000
Australia

First published by The Text Publishing Company 2020
Reprinted 2020 (three times)

Book design by Chong W. H.
Cover photo by Martin Von Stoll
Index by George Thomas
Typeset by J&M Typesetters

Printed in Australia by Griffin Press, an accredited ISO/NZS 14001:2004 Environmental Management System printer.

ISBN: 9781922330352 (paperback)
ISBN: 9781925923735 (ebook)

A catalogue record for this book is available from the National Library of Australia.

This book is printed on paper certified against the Forest Stewardship Council® Standards. Griffin Press holds FSC chain-of-custody certification SGS-COC-005088. FSC promotes environmentally responsible, socially beneficial and economically viable management of the world's forests.

Contents

Introduction

AS I write, humanity stands at a fork in the road. Unless we act decisively to phase out the use of fossil fuels, global temperatures will exceed a 2°C-rise above the pre-industrial level in a few decades, and we will risk committing every living human to climatic shocks and catastrophes that will destroy our civilisation and precipitate mass extinctions.

I do not want this essay to be an obituary. But if that is not to be, we must take the road leading to self-preservation. Climatic catastrophes are already being felt and the threat will only increase in coming years. By 2030 the average increase in global temperatures will reach 1.5°C, and Earth's climate system is likely to begin to unravel. If we fail to act, dramatic destabilisation will be felt when the temperature rise breaches 2°C, sometime before 2050. Despair is not an option. Nor is selfish complacency. Instead, this is the moment to ask what you will do.

As a nation facing enormous, irreparable damage from an increasingly hostile climate, Australia is failing its people and biodiversity miserably. We are the world's largest exporter of coal and gas, and per capita one of the worst polluters on the planet. The early signs are that Australia will continue to fail. As I write, the Federal Government sees no urgency to reduce emissions, and the minister for energy and emissions reduction, Angus Taylor, has called for a 'gas-fired recovery' from the economic impacts of the COVID-19 pandemic. He is echoed by the National COVID-19 Commission Advisory Board head Neville Power, a long-term mining industry executive who is talking up the role that expanding gas production can play in the economic recovery.[1] Scientists and, increasingly, ordinary citizens know that this cannot be. If the Federal Government wishes to keep Australians safe, the gas and coal must stay in the ground. We have almost no time to avoid the ultimate failure, and much to change. This declaration is a last call for rational action to protect ourselves and our children from climatic catastrophe. Fail now, and we will fail forever.

A recent research paper published in the world's leading science journal, *Nature*, documents just how imminent the danger we face is, noting that several tipping points appear to be 'dangerously close'.[2] Earth's tipping points involve aspects of the planetary system that, once crossed, will result in a new, hostile climate. The foremost concern of the researchers is the tipping point that will be triggered if Earth's ice sheets collapse. They write that the glaciers of the Amundsen Sea embayment in West Antarctica might already have passed a tipping point, and when they collapse, the rest of the West Antarctic ice sheet will follow like 'toppling dominoes', leading to three metres of sea-level rise. This, the researchers point out, has happened many times in the deep past. The Greenland ice sheet, they say, could soon follow, adding seven metres to sea-level rise. Its tipping point could lie at a temperature as low as 1.5°C above the pre-industrial average, which on current trends will be reached in around ten years' time.

Studies of the collapse of ice sheets at the end of the ice age around 10,000–15,000 years ago, reveal just how fast seas can rise as glaciers collapse. In the 1200 years between 14,700 and 13,500 years ago, the collapse of the Eurasian ice sheet led to the oceans rising twenty-five metres. At times, the seas were rising at the rate of four centimetres per year—more than ten times the rate experienced today.[3] The impact of such rates of sea-level rise on coastal cities would be catastrophic.

The climate models reveal that other tipping points lie between temperature rises of 1.5°C and 2°C, and will lead to events including die-off of the boreal and Amazon forests, destruction of coral reefs and melting of the permafrost. The coral reefs offer a particularly clear example, with 99% loss of global reefs occurring at a temperature rise of 2°C. Just where the tipping points lie for tundra and Amazon forest destruction, ocean circulation and the melting of the permafrost is unclear. But all are showing signs of instability at current warming, and scientists fear that their tipping points are below a rise of 2°C.[4]

The study authors write that: 'In our view, the clearest emergency would be if we were approaching a global cascade of tipping points that led to a new, less habitable, "hothouse" climate state. Interactions could happen through ocean and atmospheric circulation or through feedbacks that increase greenhouse-gas levels and global temperature.' They conclude that 'warming must be limited to 1.5 °C. This requires an emergency response.'[5]

The 26th Global Climate Meeting (COP 26), to be held in Glasgow in late 2021, will be our last chance as a global community to act effectively. The Glasgow climate summit was originally scheduled for December 2020, but the COVID-19 pandemic made that impossible. The rescheduling, resulting in a year's delay, underlines the strained and difficult circumstances the world finds itself in as we struggle with the existential threat of a looming climatic catastrophe. If there is a silver lining at all, it is that the COVID-19 pandemic has brought renewed purpose to government. As we look back on the first few months of 2020, we can see the origins of a new way of doing things that, if allowed to unfold, may just see us take the path at that fork in the road that leads to survival.

~

The events of early 2020 illuminate the path. As Australia's bushfire emergency reached its peak in January 2020, another emergency was escalating in China. A new respiratory disease—soon to be known as COVID-19—was spreading like wildfire in the city of Wuhan, Hubei Province. Australians heard the news that the number of cases there was doubling every few days, and that the disease had spread outside China. But few people had any idea of what was to come.

Australians and Chinese people had taken to wearing masks, but for very different reasons—Australians because of potentially deadly smoke from the climate-induced megafires, and the Chinese because of the new respiratory disease. Things that travel invisibly in the Great Aerial Ocean, as I've come to think of our atmosphere, represent a particular danger, because the atmosphere links us in the most intimate of ways.

Both carbon dioxide (CO_2, the main greenhouse gas) and the COVID-19 virus travel in the Great Aerial Ocean, unseen, accumulating silently and stealthily. Silence and invisibility have allowed both the pandemic and the climate emergency to escalate into crises that threaten our civilisation. And this has occurred before their impacts were widely felt. In the early stages of both crises, confusion abounded. Doctors were unfamiliar with the COVID-19 virus, and climate sceptics denied the symptoms of the climate crisis. But by early 2020 the cloak of invisibility was being stripped from both the pandemic and the climate emergency. In Australia, the Black Summer megafires made climate change visible and manifest while, shortly after, outbreaks of COVID-19 in many nations did the same for the virus.

The deaths of hundreds of thousands of people due to COVID-19 is a horrific catastrophe. And the pandemic is far from over. As the World Health Organization (WHO) has warned, the disease might always be with us. But it is what we do not know yet, both about the pandemic and about climate change, that is most worrying. Only careful but resolute action, guided by science, can see us navigate both perils.

Of course there are many differences between the COVID-19 pandemic and the climate emergency. One is caused by a virus, the other by a group of molecules known as the greenhouse gases. One affects individuals directly, the other the Earth system. And of course, the time scales over which the pandemic and the climate emergency develop are very different. Nevertheless, the response to the pandemic provides us with a roadmap for dealing with the climate crisis.

So far, the response of the Australian Government to the two crises has been very different. Bushfires of unprecedented size and intensity had been burning for many months by late December 2019, the height of the Black Summer fires. Experts had been warning for at least a year of the impending crisis, yet Prime Minister Scott Morrison refused to meet them, instead choosing to go on summer holidays. As the fires burned, his whereabouts were hidden from the public, though we later learned that he was in Hawaii. When asked why he had chosen that holiday destination, he explained that Hawaii was a good place from which to commence a visit to India. Selling more Australian coal, and accelerating development of the Adani coal mine, were doubtless high on the PM's agenda for the proposed visit, which was subsequently deferred.

Soon after returning to Australia, Morrison again fumbled badly. He authorised an advertisement trumpeting the actions the government was taking to deal with the bushfire crisis. But the advertisement was in fact not a government announcement, but a paid advertisement by the Liberal Party. Widely remembered as the man who brought a lump of coal into the Australian parliament in 2017 and jokingly saying that there was nothing to fear from it, the prime minister has much work ahead of him to re-establish his credibility.

And the government continued to downplay the fires. Angus Taylor, minister for energy and emissions reduction, was at the Madrid climate meeting arguing that Australia had a proud record of dealing with climate change and that it should be allowed credit for its emission reductions under the Kyoto Protocol.[6] Australia was alone in trying to claim such credits, as it was alone in being the only signatory to the Kyoto Protocol that was, over the first period, allowed to increase its emissions by 8%.

Meanwhile at home, the minister for home affairs, Peter Dutton, continued to repeat the falsehood that arsonists were to blame for the blazes, when authoritative research shows that less than 1% of the land burned in New South Wales and Victoria originated with arson.[7] There was also a constant drumbeat in social media, supported by the Nationals' Barnaby Joyce, that insufficient hazard-reduction burning, supposedly held up by 'greenies', had been responsible for the ferocity of the fires—despite experts noting that hazard-reduction burning has little or no effect on fires such as those experienced in 2019–20.[8]

I find it astonishing that at the same time this was going on, the Federal Government was responding to the growing COVID-19 threat in an exemplary manner. From the earliest moments of its awareness of the crisis, the Federal Government was heeding scientific advice and making decisions with massive economic implications with alacrity. Among its early and most important moves was the suspension, on 1 February, of arrivals of flights from China. This decision, which has had an enormous impact on the education, tourism and aviation sectors, was taken on the advice of Australia's Chief Medical Officer.

Also prescient was the Federal Government's announcement, on 27 February, that COVID-19 would become a pandemic, almost two weeks before the WHO officially declared the pandemic on 11 March. Despite these important announcements, the full scale of the threat had yet to be internalised by the prime minister, who was still initiating handshakes, making the Rural Fire Service volunteer who refused to shake his hand just a few weeks earlier look prescient.[9]

By 15 March, with the number of Australians being infected by the virus doubling every four days, the prime minister announced that he would stop shaking hands. We were then in the last possible moments where action might avert a catastrophe, and three weeks later we were living in a world of social and national isolation, economic disruption and unprecedented government action. It was also a world of real political leadership, overwhelming public buy-in, economic, social and medical innovation involving hitherto unthinkable measures, all of which allowed Australia to mount one of the most effective COVID-19 responses to the initial round of infection in the world.

In justifying the extraordinary measures his government had taken, Morrison said, 'The virus writes its own rules.' As the climate scientists have known for decades, so does CO_2. The massive decisions about our economy and behaviour that COVID-19 forced us to make in a very short time proved successful. A similarly resolute and scientifically informed approach can do the same for the climate emergency.

It may seem that the COVID-19 pandemic merited a more serious effort because it was more deadly than climate change. But this is just not so. Four hundred Australians died directly as a result of the smoky, polluted air generated by the climate-induced megafires, while thirty-four died in the fires themselves. By 25 August 2020, 517 Australians had died of COVID-19. The truth is that both crises are life threatening, and both demand a strenuous response.

The crises of 2020 gave federal and state governments a challenging job, and the best of them have risen to the challenge. When nothing threatens, governments try to make themselves relevant by fighting ideological battles, or by supporting or opposing 'hot button' issues that tend to divide people and cost governments respect. But crises test governments, and success can bring widespread and enduring respect.

The Federal Government's response to COVID-19 teaches us much about how we could, very quickly, address the climate emergency. This book outlines a plan—in effect a cure—for the climate crisis, which takes Australia's response to the pandemic as a model for climate action. It's a cure that would unfold over decades, and it's ambitious and urgent, for just as 15 March 2020 was almost the last moment when an effective response to COVID-19 was possible, so is 2021 almost the last possible year to initiate an effective plan to minimise the lethality of the climate emergency.

It's vital to understand that with both pandemics and climate we are not working to human-created deadlines. The progress of the climate emergency is dictated by the planetary system itself—melting ice, warming lands and oceans and rising concentrations of greenhouse gases. It is these inexorable forces that dictate what targets must be met, and by when, in order to avert disaster. Miss them, and no matter what we do thereafter, it is likely that Earth's positive feedback loops will trigger climatic catastrophe.

I was made Australian of the Year thirteen years ago, in 2007, in recognition of my cause—combating climate change. Since then, humans have emitted about a quarter of all the greenhouse gases ever emitted by our species. I worry about what my country will be like thirteen years hence, in 2033, when the seventy-third Australian of the Year is announced. The increasing probability that we will be in the midst of an unstoppable catastrophe is what drove me to write this book.

During twenty years of urging action to address climate change, I have always believed that a calm, non-emotional approach is most likely to succeed. But as I've watched governments ignore the experts and the long-predicted disasters such as Australia's megafires devastate the land, my view has changed. I now believe that those wanting climate action wrongly think that presenting facts will lead to a solution. Our opponents, however, are willing to mislead and sacrifice the public good in their pursuit of monetary profit. Australians fighting for a better climate future need to start taking risks and thinking big.

And now is the moment for action. As the world sought to recover from the 2009 stock-market crash, much of the economic stimulus was directed to supporting the fossil-fuel industry. As a result, global emissions rose 5%. We can't afford another recovery like that.

If we hope to survive the climate emergency, we need to fight three critical battles, and we can't afford to lose a single one. These three battles bear a striking resemblance to the battles required to prevail against COVID-19. The first and most urgent involves cutting fossil fuel use, decisively and deeply. This is akin to our initial actions in response to the pandemic—which involved containing the spread of the virus. And curiously the social isolation that was the key to success also reduced greenhouse gas emissions. To continue that success, we need to focus on increasing clean energy and decreasing fossil fuel use as the economy recovers.

The second involves minimising the damage that our nation and our planet will suffer as a result of the greenhouse gases already emitted and those we cannot avoid emitting in future. The key elements involve mounting an effective policy of adaptation and planning for any necessary geoengineering. This climate battle is akin to the virus action of ensuring that the nation has the critical-care capacity to deal with the stricken.

The third battle is to lay the foundations of a medium-to-long-term response to climate change, which will help claw back our way to climate stability. The analogy in the COVID-19 battle is the development of a vaccine—in this case, the development of a new, clean economy with the capacity to draw CO_2 out of the air at scale. Known as 'drawdown', this removal of CO_2 is an area in which Australia could take global leadership. But the search for a 'drawdown climate vaccine' cannot be used as an excuse to continue polluting.

For those reeling from years of destructive Federal Government inaction, fighting these battles must feel like climbing Mount Impossible. There's still a sense that we are stuck in a phase of political opportunism, where climate policy is a weapon to destroy the opposition, rather than a tool to save the nation. But I believe that the onset of the COVID-19 pandemic marks a turning point. Combatting the virus, we have learned how science-based action can lead us out of a crisis.

The catastrophes inflicted on the nation in 2020 have seen opposition to ambitious climate action crumble. Australia's bushfire season started early and by January had built to a crescendo, spawning megafires that brought home how high the cost of inaction on climate change can be. Then, in February, fire was followed, on the east coast of Australia, by devastating floods. As the floods started to subside in March, the COVID-19 pandemic was declared.

The COVID-19 pandemic might yet prove to be the most decisive of these three events. It has seen even the blowhards of the American and Australian far right, to varying extents, give up their bluster and to rely upon expert advice. Australia at last has a Federal Government that, faced with a national emergency, is implementing policies that would once have been considered impossible policies, cooperating with other nations, and preparing the country for a hard but unavoidable transition.

The public response to that change has been heartening. After years of bitter scepticism, Australians at last have confidence that their government is acting in their best interest. When the next destabilising climate catastrophe hits, I have no doubt that many Australians will be looking to the Federal Government to respond to the climate emergency in the same manner that it responded to the threat of COVID-19. Of course, Australia cannot close its borders to climate catastrophes. But there is a huge amount that can be done.

With polls showing that 84% of Australians are willing to take action on climate change, there is now no political excuse for inaction.[10] And businesses (which with the exception of the fossil fuel industry) are increasingly united on the need for climate action. Many are urging their own peak bodies, as well as the government, to greater efforts. Australians must insist that their Federal Government puts Australia on a war footing in the battle against climate change.

We are living at a pivotal moment in history, in which the balance of power, both political and in terms of energy, is shifting. In 2019, for the first time in the history of our species, the amount of electricity generated from clean sources (and that excludes nuclear) exceeded that generated by burning coal.[11] We have crossed an energy Rubicon, from the polluting, politically poisoned world of fossil energy to the world of universally available, abundant, clean and cheap power.

The motive force propelling the change is gaining momentum: the cost of clean energy fell so greatly over 2019 that it is now the cheapest form of energy generation in most major markets. In the face of this competition, already 42% of coal-fuelled power plants globally are running at losses. The Intergovernmental Panel on Climate Change (IPCC) says that 59% of coal power worldwide must be shut down by 2030 if the world is to have a chance of the increase in average global temperature staying under 1.5°C.[12] So already, we could do two-fifths of that job by shutting loss-making coal plants, and we could save money. It's estimated that by 2028 it will be cheaper in Australia to build new wind and solar plants than to run existing coal power plants.[13] If governments cease subsidising fossil fuels, and stop appeasing special interests, economics will win the first battle of the climate wars for us.[14]

So why aren't we already acting? The short answer is that the influence of the fossil fuel industry continues: it goes on defending its economic interests in ever more naked terms, demanding more subsidies and protection. And, so far, it has proved far more skilful than those interested in climate action in the battle for political action.

Despite the favourable economic aspects, the tasks required to avert the climate emergency are terrifyingly large. Even in the best-case scenario, in order to give ourselves a two in three (66%)

chance of stabilising global temperature rise at 1.5°C, humanity must halve its emissions over the next ten years. Never in history has any country achieved anything like that. But that should not rob us of hope: the world has never seen anything like 2020, and social tipping points, including a rejection of the use of coal, are also nearing.

With huge changes required in a short time, we are best to listen to the scientists as we prioritise action. And they tell us that we should begin with the electricity sector, by quickly closing down the dwindling number of antique coal-fired power plants in Australia, and replacing them with wind and solar.[15] We need a plan to unite the country in achieving this goal, and to simultaneously lay the groundwork for the adoption of clean transport and industry by 2040.

Despite the bluster of the Trumps of the world, it's clear that the tide is turning in the planet's favour. In 2019, global emissions of greenhouse gases from the burning of fossil fuels rose by just 0.6% relative to 2018 levels—to 33 billion tonnes.[16] And during 2019, emissions from the burning of coal for electricity dropped by 3%, the largest decline in thirty years.[17] It may be that we have reached 'peak fossil fuel use'. The response to the COVID-19 pandemic has seen emission levels fall in 2020, but we must not become complacent as the economy recovers.

There is also heartening news regarding the mining of fossil fuels. In February 2020 Canadian mining company Teck

Resources announced that it was abandoning its CA$1.3 billion expansion in the Alberta tar-sands.[18] Now, analysts are predicting that dozens of other oil sands projects won't go ahead.[19] In Australia the Norwegian energy giant Equinor has abandoned plans to search for oil in the Great Australian Bight.[20] With the oil price at a historic low, it looks as if both the search for new fossil fuel reserves and the exploitation of known reserves might be slowing. The recent announcement by BP, that it has wiped US $4.5 billion from the value of its assets, shows how deep the trouble the oil industry faces is. The writedown, which takes the asset value of the company from $17.5 billion to $13 billion, is the largest in the oil industry since the $32 billion writedown that came in the wake of BP's Deepwater Horizon disaster in 2010.[21]

As some mining companies abandon the search for new sources of fossil fuels, others are looking at how they can clean up their industrial processes. Rio Tinto, the world's second-largest mining company, recently announced that it will spend US$1 billion between 2020 and 2025 to help it meet its climate targets. The company is aiming at a 30% reduction from 2018 emissions intensity levels by 2030, along with a 15% reduction in absolute emissions.[22] Others are headed in the same direction, with major investments in wind and solar. Some are even looking for new sources of profit in the emerging clean economy, in, for example, clean hydrogen. BHP (the world's largest mining company), Anglo, Fortescue and Hatch Engineering—which have relatively few interests in fossil fuels—announced in March 2020 that they have joined forces to 'de-risk and accelerate' the production of renewable hydrogen.[23] The consortium's goal, 'to identify opportunities to develop green hydrogen technologies for the resources sector and other heavy industries', could have a transformative effect because, as we shall see, clean hydrogen could transform the entire economy.

We can and must build on these first glimmerings of hope. The costs of the clean energy transition have never been lower, nor the costs of inaction higher. Australia is poised to grasp a brighter, more prosperous and cleaner future. But delay, even by a few years, could cost us everything.

I believe that the climate cure outlined in this book contains the vital elements Australians require for success in averting the climate crisis. It charts a common-sense, rapid pathway forward and deals with the full range of consequences now upon us—both opportunities and disasters. And given our experience in tackling COVID-19, I'm sure that all are attainable in the very near future.

PART 1

The Great Australian Tragedy

CHAPTER 1

A History of Folly

AUSTRALIA is a climate paradox, for it is rich in fossil fuels, yet exceptionally vulnerable to a changing climate, and this means that we have purchased our prosperity at terrible cost. Our political struggles to reconcile these facts have become a diamantine version of the more diffuse dilemma the world faces as it deals with the climate emergency. Perhaps as a result, the actions of our politics, citizens and corporations have embodied the best and worst of all climate actions.

The dilemma the nation faces in its relationship with fossil fuels is not new. From the earliest days of European settlement Australia has been a resource economy, and exploiting those resources has always entailed risk. Perhaps the closest parallel to our current dilemma occurred in 1938 when 'Pig-iron Bob', as Robert Menzies became known, insisted on the export of iron to Japan, on the eve of World War II. Menzies, who was attorney-general in a conservative government at the time, was affronted when trade unionists at Port Kembla refused to load ships with pig iron when they learned that the cargo was destined for Japan, where they feared it would be made into weapons. At the same time, other ships in Melbourne were loading iron bound for Nazi Germany, again with the Australian Government turning a blind eye to the weapons potential of the cargo.

Menzies accused the unionists of trying to dictate Australia's foreign policy, and he forced them to load the ships with iron that, just a few years later, rained down on the Allied Forces in the form of bullets, artillery shells and torpedoes.

Parallels with the current Federal Government are striking. Both exported minerals that would come back to damage the country. In Morrison's case it is evermore coal and gas. Greenhouse gas emissions from the burning of these exports return to devastate Australia by changing our climate. And both are appeasing an enemy—in Menzies' case the Germans and Japanese, in Morrison's the fossil fuel industry.

Wars are declared, and eventually they end. Australia's war with Japan was followed by decades of increasing prosperity, allowing Menzies to govern for a record twenty-six years. Climate change will not be like that. The cost to Australia of a deteriorating climate is already extremely high, and unless action is taken rising costs will soon become unstoppable. And tragically, the limited wealth we reap from exporting fossil fuels will be paid for by forfeiting new opportunities in the clean economy, and many years of economic, social and environmental disaster.

Both Menzies and Morrison supported their anti-patriotic actions with bravado, bluster and deceit. In November 2019, with Australia's bushfire threat growing by the day, Morrison declared: 'The suggestion that…that Australia, accountable for 1.3% of the world's emissions, that the individual actions of Australia are impacting directly on specific fire events, whether it's here or anywhere else in the world, that doesn't bear up to credible scientific evidence.' He then went on to dismiss the idea that Australia 'doing any more or less' would make a difference.[1]

Australia contributed about 1.4% of all Allied Forces during World War II. Imagine if Menzies had said thatAustralia had only contributed 1.4% of the Allied Forces and that therefore its impact on the outcome of battles was trivial, that whether we had done more or less just didn't matter. Both then and now, Australia's actions matter very much indeed. If we include our exports, Australia's carbon emissions are the fifth largest of any nation. And Australia, of course, controls what it exports.

The Liberal Party has not always been like it was under Menzies and Morrison. In June 2007, in the last days of his government, Prime Minister John Howard announced that if he won the election, he would introduce an emissions trading scheme for greenhouse gases. But the Labor Party led by Kevin Rudd won the election, and the new government proposed a Carbon Pollution Reduction Scheme (CPRS) which it had been working on in opposition. The scheme would have introduced tradeable permits, paid for by polluting industries. Its ambition looked small at the time—just a 5–15% reduction in emissions from 2000 levels by 2020. But from a 2020 perspective, it looks very much better than the deal the nation was saddled with.

When the legislation enacting the CPRS was put before the Senate, the Greens refused to support it, saying that it was inadequate, and would lock in such inadequate climate action for the long term. Tragically, in the face of this opposition, by 2010 the Rudd government had abandoned this scheme for climate action. The full cost of the failure was outlined in 2008 by Treasury, which determined that the cost of the CPRS would have been far less than the cost of action taken later.[2]

Back in 2009, the Liberal Party was facing a titanic leadership struggle—between Malcolm Turnbull, who supported action on climate change, and Tony Abbott, who opposed it. Abbott won the leadership in a second attempt (by one vote). An outright climate sceptic at the time, Abbott would have a strong influence on his party's view on climate change.

Neither the Labor nor Liberal parties won a majority at the 2010 federal election, and three independents, all from regional

Australia, held the balance of power. They eventually supported the Labor Party, which once again tried to develop legislation to deal with the climate threat. In 2011 the Gillard government passed the Clean Energy Act, which mandated a price on carbon. Very quickly, emissions from polluting industries dropped by 7% while the economy continued to grow. Australians finally had proof positive that climate action could be effective, with no harm to the economy.

Despite the facts on the ground, Abbott white-anted the Clean Energy Act. Equally damaging, he stated that he would repeal the carbon price and abandon the other measures Labor had introduced, if his party were elected. The uncertainty this created in the minds of many limited investment in clean energy and energy efficiency. As 2012 came to an end, Abbott and his colleagues upped the amplitude of their campaign against climate action. The then agriculture minister Barnaby Joyce claimed that a roast lamb dinner would cost over $100.00 because of the carbon price. Yet when Treasury modelled the impact of the carbon price on a lamb roast, they found that the price impact was around $0.20.[3] Future generations of Australians will deal with such deceptions and their tragic consequences.

I was Australia's Climate Commissioner at the time of Joyce's big lamb-roast claim, and saw first-hand how effective it and the other false claims about the carbon price were. Amplified by radio hosts such as Alan Jones and Ray Hadley and elaborated on by the Murdoch press, the claims were widely taken as fact. Some Australians even believed that they would have to pay a carbon tax themselves. The truth was that, under the Gillard legislation, only the nation's largest hundred or so polluters would be liable. Costs, however, were passed onto consumers, with compensation offered to those on low incomes and least able to pay.

Fear is a great expunger of rational thought, and the Abbott-led Coalition won government in September 2013. Its first act was to abandon the Climate Commission and to take down the Commission's website, which contained factual information that many people had relied on to understand climate science. By 2014 the Abbott government had repealed the carbon price, and Australia's carbon pollution began once more to rise steeply.

Just as Abbott was destroying Australia's capacity to tackle climate change, other dark forces were at work to do the same globally. Back in 2009, I was chair of the Copenhagen Climate Council (CCC). The organisation was an alliance of business, community and climate groups built over the three years prior to the United Nations Climate Conference in Copenhagen (COP 15) to assist the Danish Government (which was chairing COP 15) to gain a successful outcome. The CCC included global businesses such as Intel and Coca Cola, and in May 2009 it had hosted the world's largest ever climate-focused business meeting. Fifteen thousand people attended the summit in Denmark, which ended on a high note, with me, on behalf of the meeting, handing the Danish prime minister a declaration—the Copenhagen Call—exhorting politicians globally to take effective action on climate change.

But even before COP 15 opened seven months later in December 2009, one thing after another started to go wrong. A very effective, but entirely fraudulent, campaign to discredit climate science, labelled Climategate by the media, based on the illegal hacking of thousands of emails by climate scientists at the University of East Anglia, was mounted. Just a few words and phrases, quoted out of context, were used to imply that climate science was a conspiracy and that scientists had manipulated data to suit their own ends.

While the story was initially promulgated by climate denialists, it was soon taken up by the mainstream media. In response, the American Association for the Advancement of Science, the American Meteorological Society and Union of Concerned Scientists supported the climate scientists, confirming that there was no substance to the accusations. Subsequently, eight investigatory committees found no evidence of wrongdoing on the part of the scientists, but by that time the damage had been done. Many of the world leaders who came to the Copenhagen meeting had had their faith in climate science shaken. Uncertain as to how much support remained for climate action in their home countries, they were in no mood to promote aggressive action on climate change.

Chance added to malice to further undermine the meeting.

Sudan had been chosen to lead the G77 (the Group of 135 developing nations) in 2009. But a vicious civil war was being fought in Sudan, and Sudan's President Omar al-Bashir became the first leader ever to be indicted for war crimes by the International Criminal Court. A warrant had been issued prior to the Copenhagen meeting, and Amnesty International called for his arrest the moment he touched Danish soil. Despite the fact that the Danish government offered al-Bashir an amnesty to attend the climate meeting, he was in no mood to cooperate, saying that taking action on climate change would be like committing genocide on the developing world.

Almost everyone involved in the Copenhagen meeting had great hopes that the Chinese would play a constructive role, and when Chinese Premier Wen Jiabao arrived with a retinue of 800, expectations ran high. But, unexpectedly, Tuvalu, which has a population 11,000 and is a member of the G77, demanded legally binding emissions cuts and accused China of being a major polluter. The Chinese were entirely taken by surprise and two days were spent discussing Tuvalu's protocol. Wen, perplexed and insulted, retreated from the meeting.

Then, just a few days before the meeting's end, Danish Prime Minister Lars Rasmussen ejected his environment minister, Connie Hildegard, from the chairmanship of the COP, and took the chair himself. Within hours the negotiations were in crisis, and it was only a motion by the British to defer deliberations for the night that staved off total collapse.

By the time COP 15 finished, overdue and in disarray, the only positive outcome was a deal brokered by President Barack Obama and the leaders of China, India, Brazil, South Africa and Mexico. Known as the Copenhagen Accord, it proposed a bottom-up approach to climate action which involved individual countries setting emission targets, then pledging to honour them as part of a global deal. Although bitterly opposed by the majority of countries at the time, it would lay the foundations for the successful 2015 Paris Agreement, which is based on the self-determined targets of individual countries, which can be adjusted over time in order to reach the global ambition of keeping warming to below 1.5–2°C.

The sense of failure was near total. When journalists asked Christiana Figueres, UN executive secretary for climate change, about the prospects of a future deal at her first press conference following the debacle, she responded, 'Not in my lifetime.' I felt certain that even in the best of circumstances, there would be years of delay in addressing climate action. In fact it took six years—until 2015—to get a global agreement.

Watching the train wreck of COP 15 close-up was excruciating. I knew how much was at stake: the climate modelling revealed that unless there was an agreement on swift and effective global climate action, we faced needing to make very steep reductions in emissions later in the decade. That really did feel like facing Mount Impossible. And it was equally clear that even the most stringent emissions reductions imaginable would not at that point be enough. The world would have to put increasing reliance on drawdown technologies, which at the time were barely explored, to avoid a climate emergency.

By the time the Copenhagen meeting closed in failure, I was sitting in a pub in Nyhavn—the old port area of the city—contemplating the three years of intense effort that had led to nothing, that our efforts to save humanity from the climate apocalypse had failed.

December 2009 was indeed the critical moment for climate action. Had the Copenhagen meeting resulted in a meaningful agreement, had the Greens supported Labor's Carbon Pollution Reduction Scheme, or had Turnbull held on to the Liberal leadership in 2009 and prevented Abbott's destructive actions, Australia might have played its part in ensuring that climate emergency was averted. But none of this happened. The reality was that Australia continued to increase its carbon pollution, and there was only weak and stymied action globally. At subsequent climate meetings, following the election of Tony Abbott as prime minister, Australia went on to become a wrecker of global climate action. It also became the world's largest exporter of two of the world's three fossil fuels—coal and gas. We were well down the road to catastrophe.

After 2013, Australians who continued the fight for climate action have had to look beyond the Federal Government. Within a week of Abbott abolishing the Climate Commission, with the help of Amanda McKenzie (who is now CEO), I went on to establish the Climate Council by appealing directly to the Australian public via crowd funding. All but one of the Climate Commission's commissioners joined us on the new body, as councillors or board members.

Since then the Climate Council has gone from strength to strength, and today it employs many more staff than the Climate Commission did, as it continues to provide up-to-date scientific information on climate. It has established the Climate Media Centre, which gives voice to farmers, doctors and others concerned about climate change; the Cities Power Partnership, which works with local governments to get action on climate; and the Emergency Leaders for Climate Action, all of which have been highly effective. The Climate Council has played a crucial role in creating climate awareness: today 84% of Australians want climate action—testimony to the fact that, provided with sound evidence, Australians are willing to take action.

Despite the growing desire of Australians for strong action on climate change, the Federal Government remains hostage to about twenty-five members of parliament[4] who Malcolm Turnbull described as acting 'like terrorists'.[5] And in holding the government hostage, they're continuing to hold twenty-five million Australians to ransom. But we must act, even as our leaders fail to act.

As voters we can purge parliament of the climate denialists

and win the battle for our climate security. But it will take Australians working together, with a great sense of purpose, to achieve that. Even if that does not happen, I believe that at some point the Australian Federal Parliament will vote for strong action on climate change. A growing number of politicians in all parties see the peril of failing to act quickly. They understand that change in our energy systems is inevitable, and that delaying our response could cost us the Earth

If Australians are to succeed in tackling climate change, we'll need to see with razor-sharp clarity who the enemy is. While I've talked a lot about politics and politicians in this declaration, they are not the real enemy. The real enemy is the fossil fuel industry, which is responsible for 75% of all greenhouse gases that are emitted globally. Clearly, we cannot avert a climate disaster without rapidly ending the use of fossil fuels.

The rhetoric employed by the fossil fuel industry has changed with the times. For years it and its lackeys in the media and parliament were arguing that climate change didn't exist. British climate sceptic and public speaker Christopher Monckton was taken seriously by many—until his ridiculous claims to have the cure for multiple sclerosis, influenza and herpes made him a laughing stock.

When it became impossible to deny the impacts of its pollution, the fossil fuel industry changed its line. It admitted the irrefutable—that the climate was changing—but claimed that humans were not responsible. When the human influence on the climate in turn became undeniable, it switched to the line that while we humans might be responsible, the costs involved in fixing the problem were too high. When even that was proved false, it turned to promoting despair, promulgating the idea that it's too late to address climate change and that nothing can be done anyway. These staged retreats have had an immense impact, their power to disempower and demoralise being a major factor in holding us back.

The most morally repugnant aspect of this litany of misinformation is that the fossil fuel industry has known all along just what the impact of burning fossil fuels would be. In 1982 a simple graph was produced by the world's largest oil company, Exxon. It projected the increasing concentrations of CO_2 that would result from a business-as-usual approach if it and other fossil fuel companies were allowed to pollute unchecked. The graph also correlated CO_2 concentrations with temperature. Looking at where we are in 2020, Exxon's prediction has proved to be accurate.[6]

It is shocking to think that the fossil fuel industry had begun planning its war on climate action decades ago, in the full knowledge that it would cost us our world. But that appears to have been what happened.

Learning the truth about the fossil fuel industry is confronting. We buy its products daily, and many of us have some further link with it, either through shares, superannuation or employment. A sense of outrage is mixed with guilt. How are we to respond? Christiana Figueres councils us to forgive, and move on, saying: 'We must let go of the fossil fuel dominated past without recrimination. The process of letting go is essential, and it must be intentional. The more work we do to let go of the old world and walk with confidence into the future, the stronger we'll be for what lies ahead.'[7]

Personally, I'm still struggling with the issue of forgiveness, particularly of those who continue to stymie climate action. But I can see that Figueres is right. We will need to reserve every atom of effort for the job ahead to have any chance of success in the battle to stabilise our climate. But as we do so, we must be severe on further efforts to misguide.

One of the ongoing tactics of the fossil fuel industry is to deflect blame and hide behind others. While governments around the world are influenced to varying extents by the fossil fuel industry, many do not own fossil fuel assets. Governments, including those in Australia, are incentivised by royalty payments to expand the extraction of fossil fuels, and therefore have a conflict of interest when it comes to cutting emissions. In the war on fossil fuels, the worst are collaborators.

Although governments can be part of the problem, they are not the enemy. Many Liberals, along with conservatives elsewhere in the world, are urging action on climate change, while at the state level, some Liberals, such as Matt Kean (New South Wales minister for energy and the environment), are deeply committed to climate action. Others, such as the South Australian premier Steven Marshall, are strong supporters of clean energy. The environmentally responsible Liberals need to be on board if there is to be enduring, bipartisan climate policy in Australia.

Mining is not the enemy either. Minerals mining has nothing to do with fossil fuels (though it does use a lot of subsidised diesel), and as we shift to clean sources of energy, large supplies of minerals, including lithium and cobalt, will be essential. Moreover, the minerals industry is increasingly moving to clean energy sources to power its operations, and many companies are distancing themselves from fossil fuels. There are many good reasons to oppose mining when it threatens environmental or cultural values. But that is a different issue from surviving the climate emergency. If there is to be a swift energy transition, the mining industry will need to be onside.

The workers in the coal, oil and gas industries are not the enemy either. The fossil fuel industry has treated many of its employees very poorly, neglecting health effects such as 'black lung' and other respiratory diseases, and perpetuating the falsehood that without fossil fuels the workers have no future. This cruel lie keeps communities and workers dependent on their bosses, leaving those who would otherwise act feeling disempowered.

The truth is that many fossil-fuel-dependent regions, particularly those in Queensland, can have a prosperous future in the clean-energy economy, because they possess unique assets that are indispensable in unlocking cheap, clean, reliable energy. Swift reduction in the use of fossil fuels will allow those living in these regions to go on to greater prosperity, better health and a cleaner environment.

The fossil fuel industry has also become a very successful parasite on the economy, and our bank balances. Parasites need to be undetected, or to be protected in other ways from their host, in order to persist, so it's poorly understood in Australia that subsidies to the coal, oil and gas companies cost an estimated $1198 per person every year.[8] And, just like bodily parasites, the fossil fuel industry affects our health. Each year in Australia, air pollution leads to the premature deaths of about 3000 people, and most of that pollution comes from the burning of fossil fuels.[9]

In addition to these deceptions and parasitic activities, the fossil fuel industry has masked its true nature by devoting billions of dollars to diverting effective action. And some of its strategies are very clever, playing to partial truths and heartfelt beliefs. Whenever I speak publicly many of these are raised in some shape or form. For example, I usually get a statement from the audience that the real problem we face is population growth and that climate change cannot be addressed while population continues to grow. To concentrate on population growth as a solution to the climate problem is a welcome diversion for the fossil fuel industry, for the spotlight lifts from them. This tactic works because it confuses two priorities. There are indeed excellent reasons for limiting population growth, and the best way to do that is by improving the economic wellbeing and autonomy of the poorest women in the world. Limiting population growth this way will bring enormous benefits to the poorest families on Earth. But it alone will not solve the climate crisis—indeed it is likely to cause an increase in energy demand. Unless that energy is clean, limiting population growth will not help the world stay below 2°C of warming.

Dietary issues offer another opportunity for the fossil fuels industry to shift the spotlight. I am often questioned about the impact of eating meat on the climate system. And yes, meat production is a part of the climate problem. Livestock is responsible for around 15% of global emissions, which means that there are very good reasons to eat less meat. But to make this the major focus of action would again delight the fossil fuel industry. It is responsible for 75% of emissions. So to blame the 15% and focus our efforts on livestock production lets fossil fuels off the hook, and could well end up costing us victory in the war for climate stability.

Perhaps the most often-asked question I get after speaking on climate change is 'What can I do?' Individual action is very important, both in terms of individual empowerment and in reducing emissions, so quite rightly there's a growing emphasis on what individuals can do to limit their climate impact. But individual action is only part of the story. Again, the fossil fuel industry is delighted when the spotlight shifts away from it and the regulatory measures needed to win the climate war, and onto the actions of others. So let us act as individuals, but never lose sight of the fact that the real battlefront lies in the boardrooms and marketplaces of the fossil fuel companies.

The key message here is that we must not become distracted by the seductive arguments and half-truths put about by the fossil fuel industries. In this time of climate emergency, we need to concentrate our efforts on ending the use of fossil fuels—coal, oil and gas. It's not just that they're 75% of greenhouse gas emissions but that the carbon in fossil fuels is very different from the carbon in livestock or plant matter. The carbon in a tree is part of the living carbon cycle. It was in the air before the tree grew, and it will eventually return to the atmosphere when the tree dies. But the carbon in fossil fuels has been locked away in the rocks for millions of years, and it would have stayed locked up in rocks for millions more if we had not dug it up and burned it. This means that the fossil carbon is additional to that in the living carbon cycle: it's increasing the overall stock.

CHAPTER 2

How Australian Government Policy Is Making Things Worse

FOR almost two decades, conservative leaders have argued that climate change doesn't exist, or that its impacts will be small or distant, or that it's too expensive to fix. The Morrison government does, at least, admit that there's a problem, but if its view prevails, and gas becomes the fuel of choice globally, we will lose the battle to stabilise the climate.

The Australian Government is also exacerbating climate impacts by refusing to listen to experts, distorting markets with ineffective and contradictory regulation and subsidies and, in an enormous contrast to its handling of the COVID-19 pandemic, failing to educate and prepare the community for what is coming. This creates huge uncertainty, which makes it that much more difficult for communities and businesses to plan for the future. For example, when it comes to the impacts of sea-level rise, a direct consequence of a warmed planet, we cannot afford to repeat our lack of preparedness in the face of the megafires. Urgent action by the Federal Government is required to protect vulnerable coastal communities.

It is important that the planning required to achieve this begin in earnest now. As we seek to rebuild the economy in the wake of COVID-19, we must invest in a clean, sustainable future that minimises the chance of future economic, social and environmental shocks. And we are in the last moments of being able to choose.

As I write these words, the Federal Government's minister for energy is appeasing the fossil fuel industry by promoting gas and other fossil fuels, and white-anting wind and solar power. As well as governments receiving royalties from the fossil fuel companies, many in the government stand to benefit from the fossil fuel industry once they leave politics. Climate-change-denying federal politicians are often part of the 'revolving door' pattern of Australian politics, whereby individuals enter as lobbyists for the fossil fuel industry, go on to become government ministers, and then exit politics to become directors of fossil fuel companies.

Both major parties, and even independents, have used this 'revolving door'. Labor's Martin Ferguson was minister for resources and energy. A vigorous promoter of fossil fuels as minister, after leaving politics he went on to the board of oil and gas company the BG Group, and to chair the advisory board of the Australian Petroleum Production and Exploration Association.

The Liberals' Ian MacFarlane is an example from the other side of the house. He too was the resources and energy minister, and after retirement became CEO of fossil fuel lobby group the Queensland Resources Council. The Independent Clive Palmer was more brazen. He didn't use the revolving door, but was simultaneously a senator and an owner of coal assets.

But the revolving door is not used only by politicians. Far more insidious, because it is hidden from public view, is the practice of advisors moving between companies, peak industry associations, ministerial staff and senior bureaucratic roles. At worst, this revolving door allows company insiders to write legislation in areas where they have (or will have) a vested interest.

Such people are supported by others in the parliament, including the current minister for energy and emissions reduction, Angus Taylor, and the Liberal Craig Kelly, who seems to confuse political spin and scientific information. Kelly, incidentally, became infamous for claiming, while the megafires raged in January 2020, that researchers found that all regions of Australia had shown reduced fires 'during the last fifty years', causing even his colleagues to distance themselves from him.[1]

Some politicians use fear as a political weapon. Whenever I travel to regions dependent on coal, I hear the statement: 'If we don't have coal, we've got nothing.' This despairing attitude results directly from the inaction and outright lies of the government, which downplays any idea that there might be solutions to the problems these communities face. By ignoring measures that would assist the coal-dependent communities to transition to a more prosperous future, the politicians in effect keep these communities dependent on any scraps that are thrown to them at election time. With coalmines rapidly becoming more automated and efficient, there's already enormous societal stress in many coal-dependent towns. It's a misery that the government and the fossil fuel industry need to foster if the policy of fossil-fuel-industry appeasement is to continue.

Discontinuing policies that successfully cut emissions has also been one of the hallmarks of successive conservative federal governments. One of the most successful federal emissions reduction policies was the Renewable Energy Target. Originally designed to ensure that 44,000 gigawatt hours of electricity come from renewable sources by 2020, it was reduced to 33,000 gigawatt hours on the basis that the wind and solar could not be built in time, and that to attempt to do so would push up electricity prices. The nation reached the reduced target a year ahead of schedule, and as a result electricity prices are now dropping. Proven wrong on both counts, the Federal Government seems determined to terminate the program regardless. As things stand, nothing but a sharply reducing financial incentive to those producing clean energy (including households with rooftop solar) will continue until 2030.[2]

With the Renewable Energy Target effectively capped and met, the Federal Government has only two climate policies focused directly on reducing emissions of greenhouse gases: the Climate Solutions Fund (originally known as the Emissions Reduction Fund) and the Safeguard Mechanism. The Climate Solutions Fund is a $2.55 billion investment. Only a minor portion goes to emissions reduction (and the majority of that goes to the promotion of energy efficiency). Most of the money is allocated to the drawdown of CO_2 from the atmosphere using plants. For example, the fund pays for tree-planting and changes to land use. As of 2019, the Federal Government claims that the fund had reduced emissions (mostly by drawdown) by 193 million tonnes. In comparison, the megafires of 2019–20 are estimated to have released 430 million tonnes.[3] So, during a single fire season, over twice as much carbon was emitted through bushfires as was drawn down over the decade or so that the fund has been running.[4] And of course, in our new catastrophic climate, the megafires will continue to occur.

Reducing our emissions of greenhouse gases from the burning of fossil fuels needs to be accounted for separately from drawdown. This was partially the case during the period that Australia had both a carbon price and a separate drawdown mechanism. Without a carbon price, conflating emissions reductions and drawdown is dangerous, because it allows industry to continue to pollute, and to use drawdown as an offset for their emissions. Offsetting emissions like this is no longer viable, because the climate problem is now so overwhelming that we must cut both emissions and draw down gases simultaneously. In this declaration, cutting emissions is part of the urgent, initial containment strategy to the climate crisis, while drawdown is part of a longer-term strategy—the equivalent of vaccine development.

The Safeguard Mechanism was designed to protect investments made by taxpayers in the Climate Solutions Fund (CSF). The mechanism operates by setting a pollution limit on every industry in Australia that emits more than 100,000 tonnes of CO_2 per year. Companies exceeding this limit should offset emissions or pay a penalty. But, in reality many have increased their pollution and faced no consequences, and Australia has seen a 60% growth in industrial emissions over the past fifteen years. The lion's share of this has come from the oil and gas industry. The Safeguard Mechanism is a spectacular and costly failure, while the CSF has been less effective than it could have been because it does not differentiate between carbon stored in biomass and fossil carbon.[5] Australia can do so much better.

In contrast to its climate policies, the Federal Government's approach to renewable energy has been resolute and effective in its efforts to slow the growth of large-scale wind and solar energy production. The reason is clear enough—wind and solar are the only economically viable threat to the fossil fuel industry. If we run the grid from cheap renewables, we stop burning coal, and can replace petrol-fuelled transport with zero-emissions electric models. But the politicians have a problem. Wind and, especially, solar are popular with the public. Instead of simply opposing them, politicians have found that constantly changing policy settings is an effective strategy for delaying clean energy growth.

There are many other ways that uncertainty is created including altering, changing or foreshadowing change (regardless of whether it even eventuates) in renewable energy targets. Another favourite method is changing the rules by which the energy market operates. A good example of this is the thirty-minute rule for settlement period for electricity spot prices. It was supposed to change to five minutes in 2021, which would result in lower wholesale prices for electricity. But because the change would favour clean-energy storage such as battery storage over fossil fuels, it has been delayed until 2023.[6]

Failing to fund the new interconnectors that are vital for clean energy deployment across the grid is another way of slowing the transition. Interconnectors—the great transmission lines that convey large amounts of electricity across the country—are vital for our clean future. They allow us to bring the energy of the sun, still shining strongly in Broken Hill, to evening-time Sydney, when households have stoves and lights on. They could allow us to bring the energy in powerful winds of the Great Australian Bight to Melbourne and other capitals. The interconnectors are energy arteries that enable access to the awesome natural, clean power that we, as Australians, are heirs to. Without policy certainty, investors delay their plans. So it is that by failing to set out a long-term, comprehensive energy policy, governments discourage investment in renewables and other emissions-reduction solutions.

But the impact of the government's climate policies plays out far beyond the energy sector. Some of Australia's largest investors (superannuation funds) have stopped investing in insurance companies because climate impacts are making them too risky. Yet they continue to invest in fossil fuel companies, in part because of the government's energy policies. This is the definition of madness.

Until the kiss-and-tell memoirs of the climate sceptics are published, we'll probably remain ignorant of the many ways that the Federal Government is appeasing the fossil fuel industry and holding back the transition to clean energy. Meanwhile the damage accumulates. Strong public pressure could change things. Were the likes of Angus Taylor and Craig Kelly under threat in their electorates over their stance on climate as was Tony Abbott in his, they may cease white-anting Australia's future.

CHAPTER 3

The Balance of Probabilities: It's Worse than You Think

ONE of the most depressing and terrifying jobs today is surely that of the climate scientists who are compiling the IPCC's sixth assessment report, which will be released in 2021. The research reveals that Earth's climate has deteriorated dramatically since they last reported six years ago. A preview of their work was recently published by Australian scientists Michael Grose and Julie Arblaster, in the Conversation.[1] The new generation of climate models includes a number of estimates that suggest that earlier models have underestimated the future warming caused by the greenhouse gases. Australia, the new research suggests, could warm by up to 7°C by 2100, and the world on average by up to 5.6°C if no action is taken to cut greenhouse gas pollution.

The researchers warn that 'we shouldn't jump on this piece of evidence, throw out all the others, and assume that the results from a subset of new models is the answer'.[2] But a number of new lines of evidence means that we should take them seriously as we frame our national approach to climate change.

It's clear we have now entered the climate emergency. Terrible things, such as the tenfold jump in the scale of our largest bushfires, are happening sooner than models had predicted, and other disturbing trends are accelerating earlier than expected. We must now change course quickly in a destabilising world. If we could just reset the clock, and have the last decade over again, things could be very different.

In 2011 the scientists at Australia's Climate Commission declared the decade 2010–20 as 'the critical decade' for climate action. We could foresee even then that if climate action was delayed beyond that time, we would face a climate emergency.[3] Tragically, action was manifestly inadequate, with emissions continuing to grow until 2019, and only in the first half of 2020 did they fall slightly—by 0.9%.[4]

As of early 2020, the concentration of CO_2 in the atmosphere had climbed to 418 parts per million, the highest it has been for about four million years. And 2019 saw the largest rise in atmospheric CO_2 concentrations ever recorded.[5] With the critical decade for action now behind us, we continue to emit about 55 billion tonnes of CO_2 into the atmosphere every year.[6] That's a very large—almost unimaginable—amount. And if we wanted to take out just five gigatonnes (one-eleventh of our annual total) by planting trees, we'd need to cover most of Australia in forest and let it grow vigorously for a century to average that amount per year.

As welcome as the 0.9% decline in emissions that has occurred during the early months of the COVID-19 pandemic is, it is so small that it has not registered in terms of slowing the rate of atmospheric CO_2 concentrations recorded on Mount Mauna Loa.[7] It would take a 25% reduction in our emissions to bring about a 0.2% decline in the rate that CO_2 accumulates in the atmosphere. CO_2 emissions are growing by around three parts per million per year, and it will take decades of action to turn that around.

As the emissions accumulate year on year, it is increasingly certain that we will be unable to meet even the lower limit of the Paris Agreement of 1.5°C above the pre-industrial average. Indeed, many researchers think that the greenhouse gases already in the air will, without any additional emissions, take us close to 1.5°C above the pre-industrial average. This is due to the length of time it takes CO_2 to trap its full capacity of heat and for that heat to be distributed through the atmosphere and oceans. This is particularly alarming because scientists increasingly recognise 1.5°C as the highest acceptable temperature rise, beyond which Earth will become increasingly hostile to human life. Even in the best-case scenarios, it is probable that the greenhouse gases that we will unavoidably emit over the next decade or two will most likely take us uncomfortably close to a 2°C rise.

We are already experiencing unprecedented damage from heatwaves, droughts, fires and cyclones, and the Earth's climate system is now changing so rapidly that science is struggling to keep up, and this is occurring at a rise in global temperature of about 1.1°C. Many, including the renowned climate scientist Michael Mann of Pennsylvania State University, believe that our situation is, on the balance of probabilities, likely to be much worse than even the most recent scientific discoveries suggest.[8] This, he says, is because scientists tend to be conservative in their projections of future change, and because our understanding of Earth's climate feedback loops is incomplete.

The first report of the IPCC was published in 1988, and subsequent reports have appeared about every six years. The reports incorporate contributions from mostly experts, but the IPCC itself is about more than science. Its membership includes representatives of governments of nations including Saudi Arabia, the US and Australia, and every word that the IPCC publishes must be endorsed by these representatives. As a result, IPCC reports often understate future risk, and findings can be couched in conservative or technical language.

Many aspects of previous IPCC reports, from estimates of temperature increase to sea-level rise, have consistently underestimated the danger we face. US atmospheric scientist Michael

Mann argues that even the way emissions are accounted for by the IPCC is so conservative as to be misleading. The IPCC sets the baseline for the pre-industrial era as ending at 1850. But, as Mann points out, humans had been burning coal in large volumes for a century before that; so the true pre-industrial baseline should be set at 1750. This difference counts, because using the IPCC baseline of 1850, average global temperatures have risen by about 1°C. But the rise since 1750 is 1.2°C.[9]

And, Mann argues, we humans are 110% responsible for this average global rise in temperature. Computer models reveal that natural forces such as sunspots and volcanoes would have cooled Earth slightly during the twentieth century if we had not emitted greenhouse gases. So our pollution has both cancelled out that cooling as well as adding 100% of the warming we suffer from today.[10]

Because the fossil fuel industry has foiled action for so long, the cure for the climate problem is now a difficult medicine to swallow. We must cut emissions globally by at least 7.6% per year for the next decade. No nation has ever achieved such a cut, and the only way we will succeed is if the climate response becomes an organising principle of governments worldwide.

But on the balance of probabilities we will need do even more than that. So much damage has already been done to the climate system that a cascade of severe consequences is now inevitable. We must prepare for them by optimising our adaptation to the changes that are coming. This will require financing and organising society-wide adaptation on a scale that, prior to COVID-19, was unimaginable. We must also begin the immensely challenging business of trying to understand how we can draw large volumes of greenhouse gas out of the air safely. We also need to research, and fully understand the consequences of the geoengineering projects even now being proposed to stabilise key elements of the Earth system. While large-scale deployment of such measures may be a decade or two away, I think that their use is inevitable as the Earth system destabilises. And humanity needs to understand the consequences of what is being proposed if we are to avoid the possibility that by using them we will trigger even more severe problems than those we already face. In summary, we need a comprehensive plan for survival. And Australia cannot do these this alone, but our national plan needs to be an exemplar for the world.

Australia's economy, food security and social stability are already under threat. That's because our human world is engineered for the conditions of the twentieth century. We have now left those limits behind in terms of temperature, sea level, ocean acidity and rainfall intensity, as well as for the duration and severity of droughts, bushfires and river flow. Climate change is altering the very metabolism of our planet. And as the 2019–20 bushfire season illustrates, as Earth's metabolism changes, familiar phenomena can be affected in the most extraordinary and threatening ways. This is the Anthropocene, a new era in which human influences on the Earth system rival natural forces.

CHAPTER 4

Megafire

THE nature of bushfire in Australia is changing, driven by our ever-warming and ever-drying climate. While the continent is no stranger to bushfire, Australians are now experiencing megafires that are very different from anything experienced before. Much of Australia's vegetation is to some extent fire adapted and burns readily. Some ecological communities, such as the heathlands, are in fact fire dependent, requiring relatively frequent fire to remain healthy.

In the more benign climates that characterised Australia's last few millennia, Aboriginal land management created a balance between burning and regrowth, and biodiversity prospered. Fires could still be dangerous, particularly in Australia's southeast fire-flume, where the interval between fires is normally much longer than elsewhere. This is the region of great forests of mountain ash (*Eucalyptus regnans*), the world's tallest flowering plants.

Australian historian Tom Griffiths has memorably called the environment created by their soaring, columnar trunks 'forests of ash', because not only do the trees bear the popular name 'ash' (after a European tree with similar wood), but because they are destined to be reduced to ashes. These great trees grow in conditions of ample moisture and cool temperatures for centuries. The long periods between fires allow for the accumulation of funeral pyres worth of wood around the bases of the forest giants, and great tapers of shedding bark cling to their trunks, as if designed to carry the flames upwards into canopies dense with flammable-oil-rich leaves.

In the climate Australia experienced up to 2013, an extremely hot, dry period would be experienced about once every three or four hundred years. Then, once the fallen limbs and bark accumulated at the feet of the ninety-metre-tall giants had dried out, the scene was set. On a stifling summer day, after months without rain, a hot northerly wind and lightning flashes could swiftly create a cataclysm.

It's as if the trees welcome the flames, for the ground below has long been shaded out by an understorey of rainforest that prevented the growth of young ash trees. Only immolation of the parents can eradicate the rainforest, and by now the giants are senescent—dry and hollowed. They are ready to die, and so release the seeds held in abundance in their nuts, begetting a new generation.

The cycle has been running since time immemorial. But the first such event recorded by Europeans occurred in 1851, when one eyewitness reported that:

> the air which blew down from the north resembled the breath of a furnace. A fierce wind arose, gathering strength and velocity from hour to hour, until about noon it blew with the violence of a tornado. By some inexplicable means it wrapped the whole country in a sheet of flame—fierce, awful and irresistible.[1]

Twelve people died on that Black Thursday of 1851, when the colony of Victoria was just sixteen years old. Then in 1939 the great forests of ash burned again. There were lesser blazes, but it was not until seventy years later, in 2009, that the cycle repeated in a fire that seared itself into human memory as Black Saturday. One hundred and seventy-three people perished in the most fatal bushfire the country has ever known. The Royal Commission that followed taught us much about saving lives in the face of fierce fires.

Australia's first true megafire was experienced on the east coast during the summer of 2019–20. While it was very different from the forests of ash fires, it was awful in its own particular way. I know east-coast fires well. In 1994, in Sydney, I lost a house to one, and in 2002, just north of Sydney, I fought another. But the megafires of 2019–20 were like nothing I'd seen before.

The megafires that raged for months that summer transfixed the world. They also woke Australia to the dangers the nation is sleepwalking into. It's now clear that a signature event of the Anthropocene will be the climate-change-driven megafire. Other continents had seen megafires before, with blazes in recent years devastating California, the Amazon, Southern Europe and Siberia. But Australia's megafire was like nothing previously recorded anywhere on Earth.

Fire conditions depend on local vegetation, topography and climate, and as a result megafires resemble Tolstoy's unhappy families, each of which is unhappy in its own particular way. Australia's megafire was notable for its extent: it burned 186,000 square kilometres, 21% of Australia's broad-leaf temperate forested area.[2] Typically, in dry years fires burn around 2% of broad-leaf temperate forested areas, so in a single year the Australian fire grounds grew tenfold. This is a stepwise change, which marks the beginning of a new era for fire in Australia.

Since records began globally, no fire has burned anything like 21% of a continent's forested land. On other continents, over a similar period, the figure is only 4–5%.[3] Matthias Boer, lead author on the scientific paper announcing the findings, said of the

21% burned that, 'There's just nothing like it out there.' Others say that the 21% is likely to be an underestimate, as the fires in Tasmania are yet to be added to the tally, and the fire season hadn't ended at the time the estimate was made.

As with so many climate impacts, the experts saw Australia's megafires coming, with the first scientific report warning of an increase in dangerous fires resulting from climate change being published in 1985.[4] Australia's Climate Council (for which I'm Chief Councillor) has published eleven reports over the past six years, repeatedly warning of the increasing danger of bushfires.

In April 2019 Greg Mullins, the recently retired New South Wales fire chief, briefed his fellow councillors at the Climate Council. He described the warning signs that, barring an unprecedented change in conditions, would bring a bushfire holocaust during the coming summer. He predicted that the east coast ranges would bear the brunt. The drought had been so severe, he said, that there was nothing left to burn west of the ranges.

The climatic conditions that had alarmed the fire chiefs had been developing for a long time. 2019 was the world's hottest year on record,[5] and Australia's hottest, with mean maximum temperatures across the continent almost 1.52°C above the long-term average and 0.19°C hotter than those of 2013, Australia's previous hottest year.[6] The extreme temperatures continued into 2020. On 4 January 2020, Penrith in western Sydney reached a deadly 48.9°C

(120°F), and Canberra 44.0°C, both smashing previous records.

Analysis of climate models indicate that, were natural climatic variability alone responsible, we could expect a year as hot as 2019 in Australia once every 360 years. But with the additional greenhouse gases humanity has added to the atmosphere, the odds drop to once every eight years.[7] Every day the coal-fires burn in Australia and elsewhere, we're effectively increasing the severity of the fires of tomorrow.

Hot weather is a key element in bushfires, but so is dryness. 2019 was also Australia's driest year on record, with 2018 (Australia's third hottest year) also being exceptionally dry, particularly across the nation's agricultural heartland, the Murray–Darling Basin. Never in the nation's history had there been two such low-rainfall years back to back in this region.[8] Despite flooding rains quenching the fires, as I write the drought persists across much of New South Wales.

Mullins and his colleagues argued that more resources were desperately needed if lives and property were not to be lost at a terrible scale. Mullins had gathered together a coalition of twenty-four fire chiefs and was seeking a meeting with Prime Minister Morrison. Those who had heard Mullins speak in April were shocked to learn that, as conditions worsened, the prime minister was still refusing to meet with the nation's foremost fire experts.

Any response to such a massive national catastrophe should be proportionate to the danger. If Australia were being threatened by an enemy nation the Federal Government would be doing everything within its power to recruit allies, put the economy on a war footing, and raise arms, as indeed it did with the arrival of COVID-19. But when it comes to climate-change threats, the government claims that there's nothing it can do: Australia must remain among the world's worst greenhouse-gas polluters and ongoing spoilers of attempts to strengthen a global coalition to address the problem. To do otherwise, the government says, would be 'economy wrecking'.

The 2019–20 bushfire season began two months earlier than usual, when twelve local government areas declared a 'bushfire danger period' on 1 August. Fires began taking lives and property across Queensland and New South Wales. But the worst came in late December, just as families were settling into their holidays. The period between Christmas and Australia Day is Australia's summer holidays. Offices shut and people travel to campsites and holiday houses on the golden, unspoiled beaches so characteristic of our country, to fish, barbeque and let the kids run wild. For many, it's a season that defines Australia. But in some coastal towns, day was turned to night as smoke blocked out the sun and fierce walls of flame forced thousands to abandon their possessions and seek shelter in the shallows.

In what may prove to be the biggest evacuations in the nation's history, Australia's defence forces were called on to convey thousands to safety, while volunteer firefighters and police shepherded convoys of vehicles along perilous roads and out of harm's way. Australia's only sealed east–west highway, the Eyre, was closed by fires for almost a fortnight, leading to shortages of some foods in Perth; and the nation's busiest highway, the Hume, which links Melbourne and Sydney, was cut by the fires.

The scale of the fires overwhelmed the media and bewildered the public. Until I awoke in Melbourne to thick smoke blown across Bass Strait, I was unaware that parts of Tasmania were also ablaze. Individual fires merged and created their own weather. As the smoke and economic disruption spread, more than half of all Australians were directly affected.[9] In the end the megafires took thirty-three lives, but the smoke they generated took an estimated 445 lives, and put more than 4000 people in hospital.[10]

The cities of Melbourne, Sydney and Canberra were smoke-bound for months. The air quality was frequently worse than that of Delhi or Beijing. On 6 January the smoke was so thick in Canberra that it forced the closure of government departments and the Australian National University.

Doctors, warning of an epidemic of ill health triggered by smoke exposure, recommended that on the worst days, children, the elderly and infirm remain indoors. As smoke permeated birthing rooms, mothers awoke to the reality that the first breath taken by their child would be tainted with smoke. Australians pride themselves on their environment, and as week after week people emerged from a swim at the beach besmirched by ash or struggling to breathe, the sense of outrage grew.

The smoke was ubiquitous and inescapable, and more or less acrid depending on whether it had burned buildings, forests or flesh. It insinuated itself into houses—there was no way to keep it out. Opening the front door and being confronted with its intensity became a daily reminder that something had gone dreadfully wrong in the lucky country.

There's an old British saying that fire is a good servant but a bad master. In Australia, with its unique vegetation and climate, fire can become a terrifying predator. Like all capable predators it remains hidden until it's ready to strike, so even in this fire-plagued year most Australians have not seen the flames that lurk in the forest, but many have smelt its stench.

The old certainties about our privileged position as citizens of a wealthy, developed nation have been shattered by images of Australians disembarking from warships after being evacuated from fire-ravaged towns. Footage of the convoys of fire trucks reminded us that, in this new Australia of our creation, anyone can become a climate refugee.

Another striking aspect of the 2019–20 megafires was their duration. In the past, the worst fires have typically burned for a few weeks at most. Not only did the megafires burn for months, but they harassed some of their victims for months before striking. People prepared for their onslaught, only for the flame front to turn at the last moment. Then they would prepare the buckets, hoses and other equipment again, and again and again, all the while weakening with the smoke and debilitating heat. But fire can also trap its victim—like a mouse in a corner. With fire all around, and all exits cut, your fate is in the hands of the wind.

When I looked into the eyes of this great predator's fire-hunted victims, I saw the same look I've seen in the eyes of returned soldiers, or distressed refugees. Bushfire triggers a primeval fear, one that has been traumatising our ancestors since the first ape was plucked from its family, carried away and consumed by a big cat.

Following the immediate trauma there will be years of recovery. Around a billion mammals, birds and reptiles died in the blazes, and the human cost was massive. Few things rate as highly for Australians as their homes and communities. More than 5900 buildings, including 2779 houses, were destroyed by the 2019–20 megafires, and some coastal communities had the most profitable tourist season of the year wiped out. Even where the affected have insurance, it's most likely that they are underinsured.[11] The fires will doubtless also result in new building regulations, which will add to the costs of rebuilding. With some facing the prospect of leaving or remaining homeless in stricken regional economies, all of us are less resilient to subsequent shocks, including the COVID-19 pandemic.

The fire season of 2019–20 has left deep psychological scars that will alter the way we Australians view ourselves and our country. Mental health experts warn that even as the immediate trauma fades, other impacts will emerge. Professor Lyn McCormack, a psychologist who works with communities on New South Wales south coast. She says:

> Once the disaster settles, there is evidence that many individuals experience psychosocial outcomes that linger for some time. Others experience protracted psychological problems such as posttraumatic stress and anxiety; physical health problems such as increased smoking and drinking, psychosomatic complaints, and social issues such as financial stress, perceived lack of social support. There are also problems specific to children and youth (clinginess, temper tantrums, separation anxiety, hyperactivity).[12]

The impacts of the fires will ramify for years. About a quarter of Australians have reported health effects from inhalation of bushfire smoke, and over the longer term more and more impacts will be felt. Some medical researchers fear that the consequences of smoke exposure from this year's fires will be felt for generations. Six months after the fires, concerns were arising that smoke inhalation might be responsible for a rise in premature births and underweight newborns in areas hard hit by the fires.[13]

Even the economic recovery is slow. When bushfires ravaged the small coastal community of Tathra in March 2018, sixty-nine houses were destroyed. A year on, conditions had only worsened for those affected, for their insurance provided only for twelve months accommodation, and the small number of builders available meant that little construction had got underway. Nearly two years on, many were still waiting for their houses to be completed.[14]

The loss of other infrastructure can also be disruptive. About 120 timber bridges were destroyed or severely damaged by fires in Eurobodalla/Shoalhaven/Bega Valley area, leaving many without access to their homes and properties.[15] For some people, ensuring children attend school meant camping near town or staying with friends.

The Federal Government set up a $2 billion Bushfire Recovery Fund. So far, the lion's share ($100 million) has gone to primary producers (who can claim up to $75,000 to rebuild their farms). But $76 million is to be spent on mental health, and $60 million is allocated for local governments (about $1 million each per council, with an extra $18 million for those worst affected). But by early March even the Federal Government admitted that the funding was not getting to those who needed it fast enough. $1.6 billion was still unallocated.[16]

Federal Government support seems to be largely limited to whatever local councils and social security can provide. But the biggest problem is community survival. Entire communities are on their knees, with the worst affected seeing nothing before them but a life surviving on welfare payments. To get back on their feet, communities require more flexible assistance. The Federal Government declared a national emergency, but that hasn't extended to a coherent response in the wake of the devastation.

The rains that finally put an end to the fires in the Sydney area arrived with an extraordinary intensity. In the lead-up to the deluge, the smoke thickened and the weather just kept getting hotter—just like the build-up to the wet in northern Australia. Then the rains and storms hit like a big wet, wreaking the worst flooding in thirty years. The summer of 2020 in Sydney felt more like summer in Darwin than a normal Sydney summer. And then the end of the summer brought the pandemic.

CHAPTER 5

The Decade of Consequences

TWO independent estimates of the total cost of the 2019–20 Australian megafires have been made. These estimates, which take the losses due to the Black Saturday fires of 2009 as a reference point, suggest that the overall cost of the 2019–20 fires will be about $100 billion. A third estimate, based on a methodology developed by Deloitte Access Economics, puts the cost at $230 billion.[1]

Australia's megafires illustrate how the environment, people and the economy can be affected as climatic conditions move outside the band of variability for which our institutions and infrastructure were built. There will undoubtedly be more megafires, but other climate-related dangers also threaten. The consequences for society as a whole are so profound that climate change is finally emerging as a major concern for economists, many of whom have hitherto rated climate change as a relatively small or distant risk. This was highlighted in 2020, when the World Economic Forum rated it as the largest long-term risk the world faces.[2] The *Economist*'s Intelligence Unit has made an effort to quantify the impacts, estimating in November 2019 that the global economy will be at least 3% smaller in 2050 due solely to climate change.[3] The potential negative impacts of climate change have now become so obvious that even the G20 finance chiefs, meeting in Saudi Arabia in 2020, agreed to examine the implications of 'global heating' on global financial stability.[4]

Because of its latitude, climate and unique biodiversity, Australia is suffering earlier and greater climate impacts than many other places, and it's certain that between 2020 and 2030 we will experience an acceleration. We need to anticipate what areas will be affected, and by what phenomena, for only by understanding that can we adequately plan to minimise the consequences.

These impacts will worsen. But it's important to understand the scale of impacts that climate change is having on the economy right now. The early estimates of the costs of the megafires of 2019–20 include both tangible and intangible costs. Tangible costs include things like the cost of replacing destroyed structures, as well as the economic costs of lost human lives (valued at $3.7 million each, according to the Commonwealth standard). But, strangely, tangible losses exclude livestock and crop losses and crop damage caused by smoke, which all ought to be factored in. Intangible costs include longer-term impacts such as the consequences of poor mental health, unemployment, family breakdown and suicide, which can occur years after the fire. Clearly the estimates are crude, but they provide our best current guide to the scale of the economic losses caused by the fires.

Other economic costs, caused by long-term changes to the climate, are also proving high. A study by the Australian Bureau of Agricultural Resource Economics and Sciences (ABARES), released in 2019, reveals the cost of climate impacts on the livelihoods of Australian farmers over decades. The study, by the economist Steven Hatfield-Dodds, estimates that climate change had cost Australian broadacre farmers 22% of their profit over the past twenty years. That's $18,600 per farmer, per year.[5] The damage has, of course, not been inflicted evenly. Some farmers in regions that are more vulnerable to climate change have done far worse than others.

Hatfield-Dodds' study notes that Australian farmers have been doing a fine job of adapting to the hotter, dryer conditions brought about by the changing climate, and that without their success at adapting the losses could have been far greater. Had croppers, for example, stuck with 1990s technology, their losses would by 2019 have been 49%.

It is likely that the losses experienced by Australian farmers will increase in future. The extremely dry and hot conditions prevailing across Australia in 2019 resulted in a summer crop of only half the average yield, forcing Australia to import wheat for the first time since 2007. By January 2020 a lack of irrigation water meant that the nation's rice crop was almost non-existent. Even town water supplies came under threat, with regional centres throughout western New South Wales forced to truck in drinking water. As innovative as Australian farmers are, the heating and drying trend is accelerating, and it's a fair question to ask how, if nothing is done, even the most adaptable Australian farmers will remain profitable after 2030.

As is often the case, one climate crisis compounds another. The megafires destroyed much rural infrastructure, from fences to sheds, hay stores and farmhouses. And livestock losses were high. On 6 January 2020 the federal minister for agriculture and leader of the National Party, David Littleproud, announced that more than 100,000 farmers had properties or livestock affected by the fires, with an estimated 12% of the national sheep flock and 9% of the cattle herd killed or affected.[6]

The fires also devastated grape growers, not through burning, but through smoke taint. This occurs when ripening grapes are exposed to smoke towards the end of their period of maturation, affecting the taste of wine and making it unsaleable. The entire 2019–20 grape harvest in parts of the Hunter Valley and Adelaide Hills was lost to smoke taint, and partial crop losses were experienced elsewhere. The economic impact of smoke taint in grape-growing regions was compounded by a decline in tourism due to travel restrictions and smoky conditions. Christina Tulloch, the chief executive of Tulloch Wines and president of the Hunter Valley Wine and Tourism Association, estimates that the hit to tourism alone will cost the Hunter Valley $42 million in lost income. And that was before the COVID-19 lockdown.

In an interview with the *Guardian*, Tulloch chafed at the injustice of the climate impacts: 'Our small family vineyard has been carbon neutral since 2017. We've been certified as carbon neutral by the Australian Government since March 2019. Meanwhile, other industries are free to pollute and pump greenhouse gases into the atmosphere, while agriculture and the wine industry is picking up the bill.'[7] The Hunter Valley is home to both vineyards and coal mines, and conflict between agriculture and mining in the region is already high. The megafires will doubtless accelerate the divide.

Courtesy of the greenhouse gases that we have already emitted, Australia can now expect a year as hot as 2019 every eight years.[8] And over the decade to 2030, as emissions grow, the frequency will increase. Southern and eastern Australia have been drying out for decades due to climate change, and that trend is also set to continue. While the bushfires of 2019–20 were truly exceptional in an historical context, more megafires will occur. Australians must begin planning for the megafires, and their consequences, of the future.

Australia's primary producers will have to continue to adapt to the changing climate. Well-targeted government policy could do a great deal to help, and the money in existing funding could be redirected. Yet this is not happening. The Federal Government continues to spend lavishly on drought assistance, but offers next to nothing for adapting to climate change. This is not just a semantic matter. Drought assistance includes funding for those using old-fashioned, wasteful farming techniques that damage the environment. Indeed, poor land managers are likely to need assistance earlier in a drought than their more progressive neighbours.

Many farmers are questioning how sensible drought relief is if drought is the new normal. And, to make matters worse, governments are not sticking to their own guidelines as they distribute money: almost half the councils the Morrison government announced would receive drought funding in 2019 were in fact ineligible.

Such farcical wasting of public funds must stop. In my view the Federal Government has proved itself incapable of properly administering drought funding and many other sorts of funding. The country would be far better off if a properly constituted Commission for Primary Production were established. It could be given the funds currently used for drought relief, and tasked with preparing Australia's primary producers for the challenges of the future.

The commission, which should operate at arms' length

from government, could work with communities to establish priorities for funding to establish greater resilience in the face of climate change. Funding could come from existing sources and a gradual reduction in subsidies to fossil fuels, such as the fuel levy concession miners and primary producers currently enjoy. Some politicians would of course oppose the idea, because it robs them of a powerful tool that is used to garner financial support for their party from wealthy constituents.

As I write, the Great Barrier Reef is facing what is predicted by scientists to become the 'most extensive coral bleaching ever'.[9] It will be the third such event in five years. Coral bleaching was unknown on the Great Barrier Reef prior to 1976, but bleaching events have become ever more extensive and more frequent in the twenty-first century, and the average time between bleaching events halved between 1980 and 2016.[10] The time between bleaching events is critically important, for it is during these intervals that the coral reefs recover. Less time between events means less recovery. The area of living coral on the Great Barrier Reef halved between 1985 and 2012, but between 2012 and 2020 things have become much worse, with massive back-to-back bleachings in 2016 and 2017. As coral researcher Anne Cohen says: 'Without a mix of long-term cuts in emissions and short-term innovation, there's a not-so-far-off future where coral reefs as we know them simply cease to exist.'[11] Put simply, it's unlikely that the Great Barrier Reef, in its present state, can survive more than 1.5°C of warming, and the greenhouse gases already in the air are likely to drive temperatures to this point within the decade.

The consequences of coral bleaching worldwide are terrifying, with economists estimating that damage to coral reefs could cost $1 trillion annually later this century. In Australia, the costs to Queensland's tourism industry alone of ongoing damage to the Great Barrier Reef is estimated to be around $1 billion per annum, with the loss of 10,000 jobs.[12]

But the loss of coral reefs cannot be accounted for only in dollars and jobs. The reefs are home to the greatest biodiversity in the oceans, and their destruction would reverberate throughout Earth's ecosystems, both marine and terrestrial. And the impacts on humans would be immense. Entire nations (the coral atoll nations) depend upon them for food and protection from erosion. Many consequences of the loss of coral reefs are probably not conceivable until they eventuate.

Much economic damage to Australia is locked in, due to a juggernaut that has recently picked up considerable momentum—sea-level rise. The 2015 report by some of the world's foremost authorities on climate change stated that: 'Coastal communities and industries require information on regional sea-level change' and that 'inundation maps that can be used to identify areas that are most vulnerable to rising sea levels are particularly valuable'.[13] I was Federal Coastal Commissioner in 2010–11, and helped oversee the developments of maps showing sea-level rise impacts for Australia. Some of the maps showing projected shorelines at 2100 are available at Coastal Risks Australia.[14] But they are not widely promoted by the Federal Government, nor are they as widely utilised as they could be in planning for the future.

A 2015 study by the Climate Council estimated that by 2030, sea levels will be about thirteen centimetres higher than they were on average between 1986 and 2005. But this is likely to be an underestimate, as more recent studies have shown that the rate of sea-level rise is accelerating.[15] During two months in the summer of 2019–20, 600 billion tonnes of Greenland's ice cap melted, contributing 2.2 millimetres to global sea-level rise.[16] This astonishing event is, tragically, likely to be repeated or exceeded in future summers.

The impacts on humanity of ever-rising sea levels will be worsened as rainfall intensity and the windspeeds of storms increase with the warming.[17] By 2100—just eighty years from now—unless we reduce greenhouse gas emissions substantially, the sea is likely to be about a metre higher than it is at present, and rainfall and cyclone activity much more severe than experienced normally today.[18]

The damage inflicted by sea-level rise and exacerbated by climate-induced extreme weather events, including more intense rainfall and storms, is already damaging many parts of the Australian coastline. Some of the worst affected communities are in Torres Strait, where local topography, ocean currents and warming waters are causing seas to rise twice as fast as the global average.

Saibai Island, which is just 1.7 metres above sea level, provides an informative example of the ramifications. After years of steadily increasing salt-water incursions that damaged local infrastructure and caused locals to abandon their homes, in 2018 the Australian Government spent $24.5 million building a new sea wall. Just six months later it was breached by the inexorably rising seas, forcing a partial evacuation of the affected community.

Walter Waia, an elder from Saibai Island who now lives in Brisbane, has watched as the inundation of his homeland has gathered pace. He acknowledges that 'later in life, there may only be a reef where the island used to be'.[19] Waia is urging the community to leave Saibai Island. He has a personal understanding of the enormous difficulties and the painful cultural uprooting that such a decision involves. As someone who has already moved away from his homeland, he says that he feels like a frigate bird 'always flying', but with nowhere to land. As he contemplates the impact of rising seas globally, he sees that there are 'many birds out there are just like me'.[20] By 2030 the flock of birds that cannot return home is likely to have swelled considerably, as coast-dwelling Australians threatened by rising seas join those already being displaced.

There is little media coverage of the emerging threat that rising seas already present to coastal communities. Most of the examples that have gained wider attention are those occurring near major cities. One such example is the fate of beach boxes at Mount Martha on the eastern side of Port Phillip Bay. The Victorian Government recently announced that it has given up on trying to save these iconic structures from rising seas.[21] A similar situation is playing out at Kingborough, south of Hobart, where houses are under threat. The mayor has confirmed that some residences are no longer insurable due to sea-level rise and the increased risk of flooding.[22] In Western Australia alone, between Rockingham and Dunsborough, authorities have estimated that about $1.2 billion worth of houses and other infrastructure are at risk from rising seas.[23]

In New South Wales, at Stockton beach, just north of Newcastle, an entire community is under siege from rising waters. In September 2019 Stockton Beach was closed and a childcare centre on the foreshore abandoned due to saltwater encroachment. The council has moved to install 'interim protection' for some vulnerable houses and roads, but acknowledges that it cannot protect all assets because the ongoing cost would be too high.[24] For example, the council estimated that the cost of replacing the 500,000 cubic metres of sand (50,000 truck-loads) that has already been lost to coastal erosion at Stockton Beach would be between $5 and $10 million—an estimate based on using the cheapest method of offshore sand extraction, which has its own problematic effects on the marine environment.[25]

A recent study has estimated that within eighty years, if we continue on our current polluting course, half of Earth's sandy beaches will disappear, with 12,000 kilometres of the Australian coastline affected. But the extent to which seas will rise in future will be determined by how much greenhouse gas we emit. If we cut our pollution hard and fast, many of these threatened beaches can be saved.[26] Given the cheapness of clean energy, the cost of cutting greenhouse gases is now a bargain.[27] But the economic consequences of failing to act to save our coasts will be catastrophic. A 2014 estimate puts global costs of coastal inundation by 2050, if we do not act, at A$1 trillion. In Australia alone, the cost is put at $226 billion by 2100.[28] As those living on Saibai Island should warn us, the human costs will be as stupendous as the economic ones.

CHAPTER 6

Who Does What When Governments Fail?

IT is difficult to know what to do when the government fails us. It's a long time between elections, and we don't have time to wait for the next one to deal with the climate crisis. One of the most important things that can be done is to hold people and governments to account. Federal Independent MP Zali Steggall's proposed climate bill, based on the UK's Climate Change Act, seeks to hold governments legally responsible for achieving their climate targets. It includes a requirement for a national risk assessment—effectively an analysis of who and what is at risk from climate change—a national climate adaptation program, a net zero target by 2050, and an all-important independent Climate Change Commission to hold government to account. This legislative approach has proved successful in the UK, France and Ireland, and it could move Australia away from the backslidings and failures that have marred our climate action thus far.

While becoming carbon neutral by 2050 might seem too little, too late, one virtue of the bill is that it does not prevent timelines being shortened, nor targets made more ambitious. It does, however, prevent backsliding, and will require the government to fulfil its promises, regardless of how ambitious those promises are.

At the time of writing, the Australian parliament was yet to deliberate on the bill. It would probably require a conscience vote by the members of the major parties to have it pass. I believe that the bill represents an important step forward in helping create bipartisanship on climate issues. Just imagine the world we'll be living in if climate policy remains an ideological tool with which to attack those you disagree with…

Perhaps the best way of doing that is to think about living in Jair Bolsonaro's Brazil. Deep in denial about the COVID-19 pandemic, he says he's not worried because 'Brazilians never catch anything', as a result of the filthy environment they are forced to endure, which supposedly has given them superior immunity.[1] By July 2020 Brazil had more than 2 million confirmed cases, and more than 68,000 deaths. In March 2020, Human Rights Watch, commenting on the situation, said: 'For weeks, Bolsonaro has been sabotaging the states' and his own Health Ministry's efforts to contain the spread of COVID-19 and putting the lives and health of Brazilians at grave risk.'[2] An editorial in the medical journal *Lancet* said that he must drastically change course if Brazil is to avoid disaster.[3]

Various states restricted the movements of their citizens. On 20 March Bolsonaro hit back, issuing an order stripping them of their right to do so. But four days later Brazil's Supreme Federal Court revoked that order. When he exempted churches from roles regarding social isolation, the courts struck that down too. In mid-May, Brazil was proclaimed the new COVID-19 pandemic hotspot.

Others are joining the courts in efforts to prevent a national catastrophe. When Bolsonaro lied about food shortages, for example, his own agriculture minister said the stories were false. His own health minister continued to promote desperately needed policies—until Bolsonaro sacked him. The consequences of his inaction may well catch up with Bolsonaro in future. But right now, it is only brave individuals—including health officials and workers, state governors, and researchers—who are slowing the slide to catastrophe. As of mid-June, Brazil stopped releasing data on COVID-19 in the country, but by July it had passed the UK to have the second largest number of COVID-19 cases globally (after the USA).[4]

How much easier and more effective would Brazil's response to COVID-19 have been with good national leadership? As I read of Brazil's struggles, the situation feels eerily reminiscent of the Australian Government's approach to climate change. We are fortunate that an overwhelming majority of Australians want action on climate change. But in the absence of federal leadership, can it be done? And who must do what? In Australia it is the actions of state governments, councils, researchers, entrepreneurs and financiers who understand the climate problem that are currently slowing our slide to disaster.

Among the most important entities are the state governments. The ACT was the first state or territory to eliminate fossil fuels for electricity generation. And Tasmania is on track to be there by 2022, and has now set a 200% renewable energy target by 2040, with the additional clean energy to be used to produce hydrogen.[5] South Australia is also on track to be powered solely by renewables by the 2030s.[6] These jurisdictions show what can be done in Australia if there's a political will, and successive governments stick with a plan.

Some of the larger states are catching up fast. New South Wales has recently gone from being one of the worst performers to among the best. The ten-year 'Stage 1' of the Berejiklian government's plan to reach net zero emissions by 2050 places priority on the uptake of electric vehicles. It will change building codes to make it cheaper and easier to install electric charging points, encourage the uptake of electric vehicles by fleets, and change licencing and parking regulations to encourage their uptake.[7] New South Wales is now leading the nation on electric vehicle uptake. If the states worked together to pursue the most ambitious targets and programs, Australia could do its bit to solve the climate problem.

Local councils are also doing a huge amount. Thirty-four mayors and councillors attended the inaugural lunch of the Climate Council's Cities Power Partnership (CCP) at Parliament House, Canberra. I listened with interest as one after another described the projects they were focusing on. The breadth of projects was astonishing, from promoting bulk buys of solar panels for disadvantaged residents to making low-carbon road surfaces at local plants. Many were planting trees, assisting with energy efficiency measures, or converting waste to energy. As they started discussing how they might collaborate, I felt that the CCP was already playing a vital role, just by bringing people together.

Since that first meeting the initiative has grown hugely. Now, more than 120 local governments are collaborating, representing half of all Australians.[8] Each council has taken on four actions, and a mentoring role with two other councils. Australia's local councils have become powerhouses of innovative climate solutions. I well remember one question asked on that day of our first lunch: 'We're sitting in the Federal Parliament. It has billions to spend. So why is it that your small not-for-profit organisation is doing this, rather that our federal politicians?'

It is not just Australia's local councils that are leading. Individual Australian households lead the world in producing clean energy. More than two million households—21% of the nation's total—have now installed solar panels.[9] That means that at the end of 2019 Australians had added 2.13 gigawatts of clean electricity generation capacity.[10] This, of course, was supported by the Federal Government's Renewable Energy Target. But it wouldn't have happened without Australians paying good money for their rooftop solar panels.

Movements aimed at building momentum, such as the school strikes for climate action, will doubtless continue. In September 2019 hundreds of thousands marched during the school climate strikes. Greta Thunberg's first personal climate strike had taken place just a year earlier. In Australia the crowds were unprecedented, as was their passion.

Demonstrations have had a limited impact on the Federal Government, but now people are also organising in different ways. Extinction Rebellion, an organisation just two years old, is one of the potentially more potent. Its members are committed to breaking the law peacefully. They have held numerous actions globally, including in Australia. Part of their power lies in the fact that they keep reminding the police, courts and politicians that their actions are being taken to save everybody's children, not just their own.

But among the most potent of all actions has been those taken by traditionally conservative voters who are sick of being held to ransom by the climate deniers in parliament. Independents like Zali Steggall have run for a seat in Federal Parliament and won (with an astonishing 58% of the vote). More Independents supportive of climate action would do a lot to shift our politics in the right direction.

Electing more pro-climate-action independents to the Federal Parliament will not be easy. They will be fiercely opposed by the major political parties, which have hundreds of millions to spend at elections. The budgets of independents are, by comparison, very small. But imagine what Australia could be like if the Liberal and National Parties were forced to rid themselves of the denialists because they were being challenged by independents, and so were once more able to implement rational, enduring energy and climate policies?

We are at a moment in time where this might be possible. The Labor and Liberal parties each has about the same number of members as the Hawthorn Football Club, and 'stacking' branches dilutes even that. The overall result is that a tiny, self-selected proportion of Australia's population gets to choose the candidates we vote for. One result has been to expose the Liberals in particular to highjack by climate deniers. Routing Abbott was one victory in the fight for climate action, but other deniers remain influential. Their modus operandi, as former Liberal prime minister Malcolm Turnbull says, is that of terrorists threatening to blow the place up if they don't get their way.

I believe that one of our greatest problems in creating change is that we've become used to living with governments that don't serve our interests. Many people are rightly cynical, apathetic and disengaged from politics. That's exactly where the climate deniers would like us to be. But if party members let them run the show, the nation will continue heading into catastrophe.

The Federal Government's response to the COVID-19 pandemic makes it clear that governments have the capacity to act decisively and effectively on imminent threats including climate change, but there is a lack of political will. New people will have to step up and join those who have been persevering in pushing for climate action for years. With enough momentum we can embark on the cure for this most wicked of problems.

PART 2

The Three-Part Cure

CHAPTER 7

Stemming the Spread

THE COVID-19 pandemic has taught us that the first step in dealing with an emergency is to stop the spread of its cause. In climate terms, that means cutting fossil fuel use hard and fast. We have delayed action for so long that this is now urgent. Unless we start rapid decarbonisation now, no amount of future research into solving climate change will succeed, for we will trigger Earth's tipping points, and find ourselves in the position of Lewis Carroll's Red Queen, who is forever running just to stay in the same place, or begin slipping backwards.

Looking at the fossil-fuel lobby and what it plans for the future, it is clear that we are in for a very tough battle. The front page of the March 2020 issue of the *Australian Mining Review*, which was released as the last flames of Australia's climate-fuelled megafires were being extinguished, leads with the headline: 'KING COAL: With help from the Federal Government, Australia's Coal Industry Is Headed for a Cleaner, More Productive Future'.[1] This is a declaration of war by a morally bankrupt enemy that senses advantage in the turmoil created by COVID-19 and is positioning itself to promote a fossil-fuelled recovery. And that, it will argue, demands winding back of environmental protections and an increase in subsidies for fossil fuels. Globally, the fossil fuel industries are positioning themselves as the saviours of the post-COVID-19 economy.

Make no mistake. We are now embroiled in a fight to the death of our planet as we know it. To survive, Australia and other nations must commit themselves to a great energy transformation. There has never been a better time to do it, because renewable energy is now the cheapest new energy source in Australia, and by using it we can lay the basis for a new, prosperous economy based on manufacturing and minerals processing. At a minimum we must begin now, with the reduction of Australia's greenhouse gas emissions by at least 8% per year, every year, until 2030. This great, immediate task must first focus on the nation's electricity supply. The investment required in moving away from fossil fuels will be moderately large—particularly the costs in helping coal-dependent communities to move on to new industries and a better future. But the cost of doing nothing will be far greater.

Lord Stern's 2005 report on the UK economy revealed that the cost of inaction was at least five times greater than taking effective action.[2] And since then, as the global situation has deteriorated, the cost of inaction has grown substantially, while the costs of action (because of reductions in the cost of renewables) have reduced.

One of the paradoxes of the COVID-19 pandemic is that, contrary to the situation in many other countries, in Australia, it hardly dented our appetite for coal-fired power. In April, demand for coal-fired electricity dropped by only about 1%, so COVID-19 itself is not directly cutting emissions from electricity. The dozen or so remaining coal-powered plants that contribute the lion's share of pollution from electricity generation in this country are inefficient, expensive and failing, yet propped up by subsidy and protected in myriad complex ways. It will take a direct economic or political assault on them for a change to occur.

The need to renew Australia's electricity supply is urgent on economic and energy-security grounds, as well as environmental grounds, for our electricity is more expensive than it needs to be, and the danger of blackouts caused by failing, antique coal plants increases by the year. Renewal of the electricity supply using wind and solar will result in falling prices for consumers, because they are now the cheapest new sources of electricity generation available in Australia.[3]

Even at current rates of wind and solar penetration,

electricity prices are set to plummet. In some regions such as southeast Queensland, the Australian Energy Market Commission is tipping drops in the price of electricity of as much as 20% over the next year or two. On average, it claims, Australians are likely to be paying $97 per year less for their electricity by 2021–22, courtesy of new wind and solar plants.[4]

The problem of old failing coal plants has been with us for years. During 2018, one coal-fired power plant broke down every three days somewhere in Australia. Surprisingly, Australia's newest coal generators (so-called HELE plants) are failing just as often as the antiques. A major source of failure is extreme heat, and solar plants are increasingly filling in when the coal plants are out of action. The least reliable of all are Victoria's brown-coal-fired plants. They fail so often that their unreliability endangers national energy security.[5]

The denialists have misled us about the price and reliability of electricity. In September 2016 when a tornado damaged twenty-three power pylons on major interconnectors in South Australia, the deputy prime minister Barnaby Joyce and the prime minister Malcolm Turnbull ignored the weather that caused the blackout and blamed wind turbines. Neither has apologised to the 850,000 South Australians left without power by a natural disaster.

A few potent lies propagated by the sceptics continue to hold us back. One, told over and over by the fossil fuel industry, is that natural gas is a transition fuel to a cleaner future. The concept is being spread by those who have invested $80 billion in gas projects over recent years in Australia and want to protect that financial investment.

At its heart, the idea relies on the common misunderstanding that as we increase the supply of intermittent energy (wind and solar) so must we burn more gas when wind and solar energy is not available. But Australia's energy history reveals how false that is. In the parts of Australia that have already transitioned to clean energy sources, there has been no conspicuous uptick in the use of gas. The real test case in this regard is South Australia, which has gone further than any other state in reducing its use of fossil fuels.

Between 2007 and 2019 South Australia shut down all of its coal-fired power plants, reduced its dependence on imported coal-based power and built an export industry based on wind and solar power. Outside the ACT (which runs on 100% renewables), it has the greatest percentage of wind and solar in the nation, and its energy mix demonstrates that we don't need to burn more gas to transition to clean energy. In 2007 South Australia was using 8.1 terrawatt hours (TWH) of gas, but by 2019 it was using just 7.2 TWH.[6] Over that period renewables grew from 0.7% of net demand to 54.3%.[7]

One of the most remarkable innovations is South Australia's 'big battery', built in 100 days and installed in 2017 by Tesla. It stores energy when it's cheap, and releases it when it's expensive, undercutting the more expensive electricity generated by gas turbines. Poo-pooed by climate denialists as an electric version of the Big Banana tourist attraction, during its first year of operation, it saved electricity consumers in South Australia more than $50 million. So successful was the innovation that there are now three big batteries in Australia, and the capacity of the original is to be increased by 50%.[8]

In 2020, the ACT became the first major jurisdiction in Australia to run its electricity supply on 100% renewable energy. In doing so, it cut its emissions by 40%.[9] The ACT has also legislated a roadmap to achieve net-zero emissions by 2045. Using a carbon-budget approach, Professor Will Steffen has estimated that if everyone had followed the ACT example, we could limit temperature rise to 1.7° or 1.8°C, within the Paris target range. It's odd to think of the climate sceptics in Parliament House, arguing that renewables can't run the grid, when the lights they stand under are powered via the ACT's commitment to 100% renewable energy. But the ACT also gives the lie to gas as a transition fuel. Evoenergy, the ACT's gas utility, has announced that it will no longer be connecting houses to the gas network, and that it will look for a supply of a clean gas substitute for its remaining customers.[10]

Another potent misconception is that Australia needs baseload power if its electricity needs are to be met cheaply and reliably. Baseload really is a Soviet-era command and control concept that describes an electricity supply network that is vanishing, or has vanished, globally. Today, Australia's power demands are very volatile and variable, except in a few continuous-process industries like oil refining, aluminium smelting and petrochemicals. As a result, Australia needs more flexibility to match the volatility and variability of demand. Fortunately, renewable power from wind and solar, coupled with storage, and consumer-demand management, can provide all the flexibility and variability that's needed, as well as combine to provide the 24/7 power needs of large industry.

Modern electricity grids have many electricity suppliers of various size and flexibility, from households with solar panels to wind and solar farms, as well as stores of electricity in the form of water in hydroelectric dams, and of course batteries also play a role. These 'assets' are controlled by smart grid technology, to provide a stable, continuous supply of electricity. Modern grids with distributed suppliers are able to supply cheaper, more reliable and far cleaner electricity than old baseload grids.

An even greater misconception is the need for an 'orderly transition' to clean energy. An 'orderly transition' is inevitably seen as a long transition by the fossil fuel industry, giving it time to recoup costs. The trouble with an orderly transition of this type is that it completely ignores the extreme urgency we now face in the climate emergency. In short, the cost of such an orderly transition will be chaotic disorder in the climate system.

Australia's Chief Scientist, Alan Finkel, has an important role to play in guiding Australia to an emissions-free future, but on 12 February 2020 he addressed the Press Club in Canberra using the term 'orderly transition' four times.[11] He was referring to a scenario in which fossil-fuel-dependent industries are given time to adjust or adapt, and in which gas (a fossil fuel) will provide the transition to a cleaner future.

The Business Council of Australia (BCA) would also like to see a long (orderly) transition, describing Labor's plan for a 2030 emissions reduction target of 45% as 'economy-wrecking'.[12] Australian billionaire Mike Cannon-Brookes responded by describing the BCA as 'a strongly regressive force' on climate issues.[13] As the costs of inaction accumulate, such arguments from the likes of the BCA are losing credibility and being revealed as self-serving for the companies they represent. It's becoming evident that prolonging the use of fossil fuels will be both economy- and nation-wrecking because nature isn't cooperating, nor are the laws of physics. If we had started with a strong emissions-reduction program two decades ago, we might have been able to progress with a more gradual transition over those years. But because, with a couple of exceptions, we've wasted twenty years, today we just don't have the time. We need an urgent, fast-track orderly transition starting today. We continued to pollute for two decades, and now climate change is making the future look less orderly by the month. Overall, the most orderly transition now available to us is one where strong government policy moves the nation at top speed away from fossil fuels.

But as some corporations are destroyed, many new ones will

be created. Respected Australian economist Ross Garnaut says of Australia that 'no other developed country has a comparable opportunity for large-scale, firm, zero-emissions power, supplied at low cost'.[14] So great is that opportunity that it's entirely feasible that, if we set our minds to it, by 2030 Australia's electricity supply could come almost entirely from renewable sources.

A recent simulation by the renewable energy development company Windlab, presented at ANU's 100% renewable workshop, and based on an analysis of three years of data from Australia's national electricity market, shows precisely how this might be achieved. Once energy storage and Snowy Hydro 2.0 are in place, the study claims that 96% of our electricity can come from renewables.[15] The key investments required to do this are:

1. An increase in wind energy from the current 7 gigawatts to 38 gigawatts, with substantial investments in north Queensland (where the wind is often blowing when there's no wind elsewhere).
2. An increase in electricity from solar farms from the current 2.7 gigawatts to 16 gigawatts.
3. Growth in roof-top solar from its current 9.5 gigawatts to 35 gigawatts.
4. An 800-megawatt transmission line linking South Australia to New South Wales, along with smaller lines linking Tasmania to Victoria, and some smaller transmission lines in north Queensland.

Of course, this is only one of many possible pathways to clean electricity generation in Australia. Many others have been proposed, and while they differ in detail, all agree that we can switch the bulk of our electricity generation away from fossil fuels swiftly and cost effectively. All require large investments in regional infrastructure. Can you imagine the boost to the Australian economy of creating five times as many wind and solar farms as already exist, all generating cheap electricity? And other, hidden costs would decline. A breakthrough study in the US revealed that when coal power plants in Kentucky closed, the incidence of asthma requiring inhaler use in the surrounding communities dropped quickly by 17%. And things kept getting better with a further 2% improvement every year thereafter. The cost savings were substantial, with 400 fewer hospital admissions per year, in a population of 700,000.[16]

Renewable energy provides cheap electricity, better health and will help avert the growing climate emergency. Yet it seems that this is not enough to propel the Federal Government into action. If it needs more evidence, perhaps it will be persuaded by Ross Garnaut's recent book *Superpower*, which argues that enormous new opportunities to make money and stimulate our economy will open once we build our renewable energy base.[17] But, as he points out, we are in a race with other nations and, regardless of our natural advantages, unless we start building our renewable-energy infrastructure fast, that opportunity will pass us by because others will beat us to market.

It was only a decade ago that some so-called 'experts' were arguing that renewables could not contribute more than 20% of electricity to the grid without destabilising it. How wrong could they be! Today, at times, South Australia runs on 100% renewable energy, while the ACT does this all the time. If Australians were to set a goal of 100% renewable energy by 2030, we will achieve it. Everything we need by way of technologies has now been tested or is already in place. What is required now is the planning and investments to build the infrastructure that will create and deliver the clean electricity.

Federal leadership could make achieving this goal no more challenging than the provision of other large infrastructure projects. The vast and evermore cost-effective rollouts of wind and solar around the country prove that we can do it. It's the 'revolving door' brigade and the fossil fuel lobbyists in the Federal Parliament that continue to frustrate progress.

Recently people have begun to ask why the states aren't just going it alone in pushing the clean energy transition, and in fact much is being achieved this way. The states don't need the Federal Government to coordinate an all-states effort to switch to a clean, renewable-powered grid. Admittedly it would be much easier with Federal Government buy-in. But if we must usher in a new age of clean prosperity for Australia without the feds, we should pressure our state premiers to do so.

CHAPTER 8

An Australian Coal Compromise

A critical element in the energy transition involves social justice. As the transition is made, we must ensure that no Australians are left behind.

One of the most daunting aspects of making the transition to 100% renewables is the impact that moving away from fossil fuels could have on regions whose economies are heavily dependent on coal mining. Places like the Latrobe Valley in Victoria, the Hunter Valley in New South Wales, and the great coal basins in Queensland.

I have visited the coal towns, both as Climate Commissioner and as a private citizen, and have seen the empty shops and other signs of communities under stress. I will never forget one meeting in Gladstone when I was Climate Commissioner. I had just given an overview of climate science, including an explanation of the causes of drought, when a giant of a man, the coal-dust staining his skin, raised his hand. He explained that he hadn't always been a coalminer. He had been a farmer, but the drought, which he now knew was related to climate change, had sent him broke. He had two young daughters, he added, and needed a wage. So he had gone to work in the coalmines. 'Can you tell me', he asked, 'am I doing the right thing?' I was heartbroken for the man, barely able to respond at first. But finally I said that we all must put our family's needs first. I ached that he and his family had been so let down. Even then there were strong alternatives for regional economic development. He and his community deserved so much better, and they could have received it with the appropriate government policies.

The coal towns are the battlegrounds where, all too often, political action on combating climate change has been stymied. And as long as the climate debate occurs in an environment of disadvantage, and is driven by fear, nothing will change. That's just how the climate denialists want it.

More investment in fossil fuels will not bring enduring prosperity to the coalmining regions. With automation and ever-increasing production efficiencies, jobs continue to decline even as the volume of coal mined goes up. And the toll on workers' health remains truly horrific, with entirely avoidable yet deadly diseases such as black lung continuing to spread, and poor air quality affecting entire communities.[1] To date, government has done little or nothing, beyond a few election-time giveaways, to improve life for coalminers and their communities.

The greatest human tragedy imaginable is the theft of hope. Many people living in these areas believe that if they lose coal, they will have nothing. The first step in creating hope is to demonstrate to those working in the fossil fuel industry that the majority of Australians stand united in their determination that the clean energy transition will leave nobody behind, and that realistic plans, which can begin with investments right now, exist to bring prosperity based on clean energy development and other initiatives to the coalmining areas.

What I'm proposing is the opposite of the infamous 'green convoy' led by Bob Brown, which visited the Galilee Basin prior to the election that returned Morrison to power. It gave the impression of a green invasion that pitted environmentalists against coalmining communities. Environmentalists instead need to listen to the communities in the coal and gas basins, to reassure them that they and their economic welfare really count, and to begin exploring together ways to ensure a better future.

Australia is not the first nation to face the challenge of supporting people dependent on industries in eclipse. The United Kingdom under Prime Minister Margaret Thatcher saw entire regions wallow in deep poverty as the coal and associated industries that supported them shut down. But in more recent times in Germany, a much better way has been found to move away from coal—one that brings better health, a cleaner environment, and enduring prosperity to regions that once were entirely dependent on coal. There is no better model for Australia to follow, as we decarbonise the economy, than what has become known as Germany's Coal Compromise.

Germany's Coal Compromise is part of a very challenging energy transition. Following the tragic meltdown of the nuclear power plants at Fukushima in Japan in 2011, Germany announced that it would eliminate all energy generation from nuclear power. It had already embarked on a process to eliminate fossil fuels. As late as 1990 coal provided nearly 60% of Germany's electricity generation, the nation's seventeen nuclear power plants about 25%, and wind and solar a mere 4%.[2] For a heavily industrialised nation with poor solar resources, phasing out coal and nuclear power is a herculean task.

The *Energiewende* (energy turning), as the Germans refer to their transition to clean energy, has involved the buy-in of communities, government and industry. The deployment of solar, in particular, has been heavily subsidised, and the negotiations around the closure of coal and nuclear have been difficult and complex. But overall the *Energiewende* has been a great success, with renewables now supplying more than 40% of Germany's energy, with coal less than 30%, and nuclear about 12%.[3]

This has been achieved while the German economy has grown, and with the German electricity grid maintaining the highest levels of reliability. And plans to increase wind and solar's share of the energy mix to 60% by 2030 look assured of success. With this success German coalmines and coal power plants are closing, in a trend that will only accelerate. And that is where the 'Coal Compromise' policy comes into play.

The Coal Compromise, which will a include a full shut-down of all German coalmines and coal power plants by 2038, was agreed in January 2019. It will ensure that, despite the mine closures, workers will soon start being re-hired as Germany's coal-dependent regions become centres of a diversified economy that includes clean energy hubs, tourism and environmental remediation.

Clean-energy generation plants, such as wind and solar, are being constructed throughout the old coal zones, and old infrastructure is being repurposed. Some underground mines, for example, are being turned into museums, while others are being utilised to store energy in the form of pumped hydro (which involves storing water to turn turbines). Elsewhere, open-cut mines are being flooded to create tourist attractions, and much-needed mine-site remediation projects are underway. As a result of this refocusing, not a single coal worker will lose their job as Germany transitions to clean energy. The cost is substantial, including an A$7 billion compensation payout to corporations, and a A$65 billion investment over coming decades in environmental remediation and the development of new industries. But the benefits, from improved community health, to cheaper energy and the savings from ongoing subsidies to the dying industries, are substantial.[4]

The belief that Australia's coal regions will lose out as we combat climate change is probably the greatest political barrier Australians face as we seek to win the climate war. I believe that if one of the major political parties announced an Australian Coal Compromise, it could change our politics by allaying the fears of many vulnerable communities that they will be left behind as Australia's own energy transition progresses.

Australia is even better placed than Germany to forge a Coal Compromise. Coal regions from Victoria's Latrobe Valley to the Hunter Valley in New South Wales have good wind and solar resources. In the Latrobe Valley, for example, a huge wind farm is being planned in an area of plantation forest. It will be the first forest-based wind farm in Australia (though they are common in Europe).[5] The farm is particularly valuable because its production is counter-cyclical, meaning that the wind is often blowing over the Latrobe Valley when it's not blowing at Victoria's other wind farms, so the electricity it generates will command a higher price, as well as contributing to a more continuous and reliable supply of electricity.

Being close to major capital cities, the Latrobe and Hunter valleys also provide other opportunities for economic development that demand access to large areas of land at low prices to satisfy demand created in the nearby cities. These opportunities range from industry to agriculture, tourism and education. The situation in Queensland is admittedly different, because the coalfields there are so far from major population centres. What north and central Queensland does have, however, is an extraordinary abundance of solar resources and counter-cyclical wind capacity that combined could make Queensland the nation's clean-energy powerhouse.

Counter-cyclical winds are particularly important to Australia's clean energy future. Wind turbines can generate huge amounts of electricity at very low cost, but when the wind isn't blowing where the wind farms are we have to fall back on other, often more expensive sources of electricity generation. But in north Queensland, the wind is often blowing strongly when it's calm over the rest of the nation, providing the potential for wind-generated power for other parts of the nation. This means that wind farms in north Queensland are particularly valuable. Queensland's traditional coal regions can flourish on the back of the abundant clean energy that wind turbines can generate. And cheap clean energy will open the door to clean new minerals-processing industries and provide clean power for hydrogen production that will begin to replace gas. But none of this will happen while those in the fossil fuel regions see greenies as the enemy. Australia's coal workers need to know that the whole nation stands committed to a just transition that will leave them and their families and communities better off.[6]

CHAPTER 9

Clean Industrial Energy

INDUSTRIAL energy is consumed in large volumes, in many cases to produce very high temperatures. It supplies energy-hungry operations such as minerals and gas processing, chemical manufacture, glass-making and cement production. The Australian Government's Safeguard Mechanism was meant to stem rises in emissions from industrial energy use, but it has failed dismally: between 2005 and 2020 Australia's emissions of greenhouse gases from industrial sources rose by 60%. If nothing changes, these emissions are projected to grow by 110%, relative to 2005, by 2030.[1]

A large part of the increase in Australia's industrial greenhouse gas emissions has come from the processing of natural gas.

Gas needs to be compressed to transport it to the massive refrigeration plants that turn it into an extremely cold liquid before it can be exported. To achieve these steps, some of the gas extracted from underground is burnt in gas turbines to power the huge transport compressors and to drive the massive refrigeration plants. It takes about 20% of the gas extracted from the ground to compress and liquefy the rest for export. Some coal-seam gas producers in Queensland use electricity from the state's coal-dependent grid to power the compressors. Either way, compressing natural gas and turning it into a super-cold liquid for export results in a lot of CO_2 entering the atmosphere. And that is even before the gas product itself is burnt. Australia is now the largest exporter of natural gas in the world, with major production plants in Queensland, the Northern Territory and Western Australia. The increasing emissions coming from gas compression and liquefaction processes in these plants has cancelled out all the reductions made in all other areas under Australia's Renewable Energy Target.[2]

The catastrophic failure of the Safeguard Mechanism has come about because the gas industry, along with other major polluters, has been granted special concessions that have allowed it to continue to pollute, yet also avoid financial penalty. The expansion of the natural gas industry has had many consequences for the nation, but one of the most divisive has been the impact of coal-seam gas extraction on agriculture and waterways. Direct competition for water and prime agricultural land, along with the disruption to primary production that's come from a proliferation of extraction wells, has created strong opposition in regional Australia. Along with a growing awareness of the impact of climate change, it's contributing to a growing demand for clean energy in Australia's regions.

An additional problem created by gas, in terms of climate change, is fugitive emissions—gas that leaks from tapped underground coal seams, well-heads and pipelines, production plants and urban gas-delivery networks. These are of growing importance: fugitive emissions from gas in Australia have risen by 41% since 2005.[3] And because gas is made up of mostly methane (a greenhouse gas twenty-three times more potent than CO_2 over a 100-year time period), this is a serious environmental problem. Unsurprisingly, it's a problem that governments have largely ignored.

Australia's gas boom has been problematic for other reasons as well. Oil- and gas-mining companies are prominent among the one-third of large companies that pay no tax in Australia.[4] And the gas boom has pushed up the price of gas for consumers in Australia. Australian gas has traditionally been very cheap. But as a gas export business has grown, suppliers have taken the higher prices overseas, and not sold to Australia's domestic industries. Due to this international competition, and the resultant shortages of supply, many Australian businesses that depend on gas have suffered job losses.[5]

According to the chair of Australia's consumer watchdog the Australian Competition and Consumer Commission (ACCC), the campaign mounted by the gas companies to get approval for the gas boom appears to have been built on misinformation.[6] Whatever the case, the problems that the boom would cause for Australian industries was easily foreseen.

In the face of the multiple problems the gas boom is causing, the Federal Government is steadfastly looking the other way. And there is no sign that the environmental mess cause by the gas boom will be cleaned up. Yet there are some faint glimmers of hope that, despite government neglect, emissions from other industrial sources will drop, even with good prospects for growth in the industrial sector.

Hope lies in the recent shift away from fossil fuels towards renewables. One important indication of this trend can be seen in the announcement by Oz Minerals, one of Australia's big miners, in February 2020 of a major investment in wind, solar and battery storage, which will provide 80% of the power it requires to run its West Musgrave copper and nickel mine. Until very recently, the natural choice for energy generation at remote mine sites has been diesel or gas, and the reason Oz Minerals CEO Andrew Cole gave for the decision was particularly encouraging. He said that while the cost of supplying renewable energy for the project was about the same as that of gas, everyone in the mining sector needed to aspire to reduce emissions.[7]

As the price of clean energy continues to drop, other mining companies are looking at following Oz Minerals' lead. Rio Tinto, the world's second largest miner, has recently announced a $1 billion investment in renewable energy to reduce its emissions. Among the initiatives to be funded are a $100 million solar plant for its Koodaideri mine in the Pilbara and a lithium-ion battery storage system to help reduce its use of fossil fuels.[8] If the Federal Government had a strong climate policy, such initiatives could proliferate, resulting in fast and sharp emissions falls in the mining and mineral-processing sectors that would make up, to some extent, for the ongoing failure of its Safeguard Mechanism.

CHAPTER 10

Cutting off Fossil Fuel's Support Network

THE fossil fuel industry continues to survive only because it has financial backers and companies willing to insure it. Over the past few years, significant advances have been made in choking off these vital resources. But much, much more still needs to be done.

Action began with the divestment movement, which was started by American environmentalist Bill McKibben on American university campuses in 2010. By 2015, it was reportedly the fastest growing divestment movement in history.[1] Despite the speed of its growth, it has proved difficult to quantify the impact of divestment on emissions from fossil fuels. Some, including Bill Gates, argue that divestment has had zero impact, presumably because many others remain willing to move into the void and risk their capital on the polluting industries. In Gates' view, you can have a more positive impact by investing in new, clean technologies. But as one large and respected investor after another has announced its withdrawal from fossil fuels, the pressure on the remaining investors grows. At the very least, the divestment movement has brought into increasing question the social licence of the fossil fuel industry to operate.[2]

Risk and insurance are intimately linked, and as the fossil fuel industry faces increasing regulatory and other risks, the pool of insurers available to it is shrinking. Risk has to be perceived if it is to be avoided. Yet the leaders of the fossil fuel industry can be mistaken in perceiving risk. In 2007 I chaired a meeting between Ban Ki Moon (then UN secretary-general), Tony Hayward (then CEO of BP), Shai Agassi (CEO of Better Place, an electric vehicle related company) and other industrial leaders organised by the Copenhagen Climate Council. I had hoped that the attendees would agree on some measures aimed at emissions reductions because, under Lord John Brown, BP had invested significantly in solar. A robust discussion on the electrification of transport ensued, but there was no common ground. As we were walking out of the meeting I asked Tony Hayward why BP was pulling out of solar-panel manufacturing. 'Too risky,' he replied. He said that BP's major focus would be on deep-water, offshore drilling. Just months later, the Deepwater Horizon drilling rig explosion in the Gulf of Mexico caused the largest marine oil spill in history. Hayward had confused risk with familiarity, and the unrecognised risk of deep-water drilling eventually cost him his job.

Insurers do a better job at assessing risk as their business is built on understanding and pricing it. They are now stepping away from some fossil-fuel operations, for example by refusing to provide insurance to new coal projects. This is happening in Australia too. In 2019, Suncorp and QBE, both major Australian insurers, joined the major European insurers in refusing insurance to coal projects.[3] The trend is accelerating, with the world's largest insurer, Marsh, reportedly considering whether it should continue to insure coal companies at all.[4]

The withdrawal of insurers from the coal business is largely being driven by the massive financial risk facing the coal industry. A report by the UK think tank Carbon Tracker has found that more than 60% of global coal power units are generating electricity at higher costs than it could be produced by building new wind and solar plants. In free markets, and without government protection, these plants are at extreme risk of immediate closure.

The same report also found that the coal industry has plans to build another 499 gigawatts of coal power globally, at a cost of US$638 billion. Because renewables can already generate electricity more cheaply than coal in all major markets, and the proposed new coal plants must remain competitive for their life of fifty years, this money is effectively being thrown away.[5] Both China and India face particularly high levels of risk in this regard, due to their plans to expand their coal power plants.

The investor exodus from coal continues to grow. Between 1 January and mid-May 2020, thirty-two global investment companies had announced that they were divesting from coal.[6] With 60% of global coal assets unprofitable, which includes 30% of Australia's coal mines, this is hardly surprising. While the coal industry insists that the good times will return, investors are increasingly wary.

The fact that the divestment movement was started by a single person, Bill McKibben, should give us all hope that we can make a difference. Both divestment and the insurance of fossil-fuel-based industries are areas in which an individual can help change things: from agitating at shareholder meetings, to asking your financial institutions and insurance company whether they invest in or insure fossil fuels. It's clear that climate change is very much on the agenda of lenders and insurers, and each call from a customer weighs in the balance.

CHAPTER 11

Hydrogen

HYDROGEN is a gas that can provide a means of storing and transporting energy. When it is burnt in fuel cells it can be used not only to generate electricity or drive transport, but it can also be used in other chemical processes as a feedstock to create an enormous range of products, from fertiliser to e-fuels and even food. In effect, hydrogen can replace all fossil fuels. It can be produced from either renewable or dirty (CO_2-emitting) sources, and there is a titanic struggle going on in Australia at present as to whether Australia will pioneer clean or dirty hydrogen production.

Clean hydrogen is produced when electricity from wind and solar is used to break apart water molecules (H_2O), releasing hydrogen (H_2) and oxygen (O_2) in a carbon-free process. Dirty hydrogen is derived from coal and gas in a process that leaves behind a stream of CO_2. The volumes of CO_2 created when fossil fuels are used to produce hydrogen vary, but using natural gas as a feedstock in a widely used process known as 'steam reforming', for example, generates 9–12 kilograms of CO_2 for each kilogram of hydrogen produced.[1]

Australia's Chief Scientist Alan Finkel is a great believer in the hydrogen economy. 'There's a nearly A$2 trillion global market for hydrogen come 2050, assuming that we can drive the price of producing hydrogen to substantially lower than A$2 per kilogram,' he says.[2] But he thinks we need to use coal and natural gas, as well as water with wind and solar, to produce it. As he explained at the National Press Club in 2020: 'We can use coal and natural gas to split the water, and capture and permanently bury the carbon dioxide emitted along the way.'

As you might suspect, the fossil fuel industry and its backers are not opposing the development of the hydrogen economy outright. Indeed, they are promoting their role in it. The very first coal-to-hydrogen plant in the country is about to open near the Loy Yang coal-fired power plant in Victoria's Latrobe Valley.[3] For every eighty-eight kilograms of hydrogen that this plant produces, it will use two tonnes of brown coal. As coal is mostly carbon, and one tonne of carbon makes 3.66 tonnes of CO_2, the pollution stream is considerable. The hydrogen that the plant produces will then be compressed (which requires much electricity, in this case derived from burning coal) before being shipped to Japan.

Over its life this pilot plant will produce three tonnes of hydrogen and 100 tonnes of CO_2, at a cost of $500 million.[4] And while the plant is touted as being 'carbon capture ready', the CO_2 it produces will be released to the atmosphere. The larger scale plant that will follow it will only be required to demonstrate that it can actually capture and store the prodigious volumes of CO_2 produced after it goes into full commercial production. And, with the government's failed Safeguard Mechanism in mind, it is not likely that such a large commercial facility would be shut down if it is discovered that it's not economic or feasible to sequester the carbon.

The entire idea that the trial plant is 'carbon capture ready' is a dodge. From a sustainability perspective, carbon capture and storage (CCS) is the most critical element in the plan. Why not demonstrate it in the planning stage along with the rest of the 'clean' hydrogen production process? After all, without it, the plant is just another consumer of fossil fuels—and one with a stupendously high output of greenhouse gas.

I suspect that the CCS process, which involves capturing CO_2, compressing it, transporting it and injecting it into suitable geological strata, is not viable, and that the industry knows it. After all, CCS has a long and chequered history, and today is mostly used successfully when CO_2 is injected into the Earth's crust to push more oil out of depleted underground reservoirs (and then goes on to be burnt producing further CO_2 emissions). But in other instances, such as in electricity generation, CCS has been an economic failure.

The Australian Government, which has long been a supporter of CCS research and demonstration programs, which have failed or delayed efforts at reducing electricity emissions, is again showing interest, especially in relation to the gas industry. If CCS is used to bury CO_2 emitted from fossil fuels burnt to make hydrogen, it could theoretically neutralise the impact of using fossil fuels as the energy source. But no CCS process buries all of the CO_2, so some emissions will occur. Overall, the main impact of CCS, as applied to fossil fuel use, is to raise the cost of fossil fuels—the CCS equipment and operating costs must be paid for, either by the taxpayer through a subsidy to the polluting industry, or by the energy customer.

Finkel argues that burying the CO_2 produced in hydrogen production will be more cost-effective than sequestering the CO_2 from coal burning, because the 'cost of extracting the CO_2 if gas is used is low'. But even if that is true, the cost of purifying CO_2 is just one cost in a long line of costs required to sequester the prodigious volumes of CO_2 from a hydrogen production plant into secure rock storage. There is no indication that, without massive subsidies, hydrogen production with CCS will be economic in the Latrobe Valley.

Finkel also argues that we should use fossil fuels to produce hydrogen because, at some point in the future, a potential shortage of materials required to build wind and solar might push up the price of renewable energy. This is theoretically possible, but thus far the price of electricity from wind and solar has done nothing but drop, making his argument sounds like special pleading.

Finkel's arguments here also seems to gloss over some obvious points, such as the wild variability in the price of natural gas, which makes using it to produce hydrogen even more risky financially than using renewable energy. And it's an obvious truth that any increase in the price of raw materials used to build wind turbines or solar plants won't make electricity from existing turbines more expensive. Solar plants generate electricity for twenty-five years, and wind plants for fifteen, after they are built. Even if we accept Finkel's assumptions, they only indicate that gas should be used to generate hydrogen if the supposed spike in prices in renewables actually eventuates. We should not be investing in Victorian coal-to-hydrogen production until we have demonstrated cost-effective CCS in the region.

I don't understand why Finkel and the gas industry are so keen on producing hydrogen from gas. It only makes sense if some gas companies already know that their undeveloped gas deposits will never be able to be developed, and that they see hydrogen as a means of creating a new market for assets (undeveloped gas) that are on their balance sheets, and for which no commercial market currently exists.

From a climate perspective, clean hydrogen produced from wind and solar is very different from hydrogen produced using fossil fuels. The process runs electricity through water, breaking the two hydrogen atoms from the single oxygen atom in the H_2O. No greenhouse gases are emitted. I find it extraordinary that Finkel talks down renewables as a source of hydrogen, saying that 'if hydrogen is produced exclusively from solar and wind electricity, we will exacerbate the load on the renewable lanes of our energy highway'.[5] But there are already proposals underway for hydrogen plants that will not use Australia's existing 'energy highways', as the electricity transmission lines that carry electricity around the country are known. Some of these will produce hydrogen at the electricity-generating source (for example, at the wind farm), while others will rely on 'energy highways' they build themselves to carry the electricity from the renewable powerhouse to the hydrogen plant.

The Asia Renewable Energy Hub is one example of hydrogen production using a self-created energy highway. It will consist of a huge wind and solar project in the Pilbara region of Western Australia, where fifteen gigawatts of wind and solar power will be used to produce hydrogen for export to Asia. This project is positioned to operate separately from Australia's 'energy highways' without any impact on them. The Macquarie Group hopes that the $30 billion investment will reach financial closure in 2023.[6]

Another example comes from Tasmania, which has announced a $50 million plan to boost hydrogen production in that state, along with a 200% clean energy target to power its hydrogen plants. Tasmania will be generating its clean electricity and hydrogen locally, so it too will have no impact on 'energy highways'.[7] But even where Australia's energy highways have proved unable to take the growing volume of electron 'traffic', we have not traditionally sought to limit industrial expansion, but instead to enlarge the highways so they are fit for purpose. I can't understand why Finkel is downplaying clean hydrogen in these ways. His views seem to be based not on what's possible with science, engineering and economics, but on what he imagines is currently politically acceptable.

Progress on hydrogen continues. In western Sydney, gas distributor Jemena has announced plans to feed hydrogen into its gas network over the next five years. Its plans are underpinned by the purchase of a $15 million electrolyser which frees hydrogen from H_2O using electricity generated from wind and solar power.[8] In South Australia, the nation's largest electrolyser is being trialled at Tonsley Innovation District in Adelaide, while the company H_2U has plans to build facilities in Queensland and South Australia to turn wind and solar power into hydrogen and ammonia for export, along with turbines that will use hydrogen for electricity generation.[9]

Around the world there's wide recognition that Australia is

well positioned to become a clean-energy hydrogen superpower. Our extraordinary solar and wind resources, high technological competence, existing gas pipeline infrastructure, and proximity to markets in Asia give us the edge over competitors. The early hydrogen-from-renewable-energy projects show the way. But Angus Taylor and other denialist politicians want Australia's hydrogen to come from fossil fuel even though there is not yet the requisite CCS capability to make them emissions free.

These developments are only the beginning of a much larger hydrogen story—one that will challenge conventional fuel use head-on in some of its most important uses. And some of the most promising glimmers of hope come from the use of hydrogen in the steel industry. Traditionally steelmaking involves the use of the carbon in coal to take oxygen out of the iron oxide molecule, in a process known as 'reduction', to produce the iron metal (and CO_2 as a by-product). But hydrogen is a superior reductant to carbon in steelmaking and has the added benefit of producing no CO_2 by-product. It has recently been trialled by one of the world's largest steelmakers, to replace coal in this part of the steelmaking process.

In November 2019 the German steelmaking giant Thyssenkrupp announced the breakthrough: it had injected hydrogen into one of its blast furnaces and made pig iron.[10] Traditionally, it takes 300 kilograms of coke and 200 kilograms of pulverised coal to make a tonne of pig iron. Using hydrogen at this stage of the steelmaking process, the company estimates, could save 20% of the emissions involved in making steel.[11] The company has set itself a target of cutting CO_2 emissions to zero by 2050, and sees clean hydrogen as its key tool in doing so. Within five years it hopes to be operating large pig-iron-producing plants, using solely hydrogen produced using wind and solar power in the reduction process.

It's worth considering here that we are talking about 'sunny' Germany, where the best solar resource is worse than the worst solar resource in Australia (which is in southwest Tasmania). If Germany can use clean hydrogen to make steel, then Australia could do it at a canter. In terms of access to clean energy and iron ore, Germany can't hold a candle to Australia. Australia is the world's largest exporter of iron ore, and much of this ore is mined in regions with some of the finest solar resources on the planet. Yet we export raw iron ore, with no value added whatsoever. Any Australian Government worth its salt would be doing all it could to kick-start a renewable hydrogen-fuelled iron-ore-to-pig-iron industry in the Pilbara.

So good are the prospects for steel made using hydrogen that a recent report by the Grattan Institute suggests that we start the national energy transition with investments in hydrogen-produced steel, and that Australia should set out to capture 6.5% of the global steel market, which would generate $65 billion in annual export revenue and create 25,000 manufacturing jobs in

Queensland and New South Wales.[12]

We cannot afford to ignore this opportunity, for progress with clean steel is developing at near lightning speed. On 13 May 2020, a Swedish steelmaker announced the first commercial production of steel using hydrogen.[13] And as I write, China's largest steelmaker has decided to fast track its own hydrogen-based steel project. It hopes to be making steel without coal in fifteen years. Currently most of China's coking coal comes from Australia, and our nation needs to think about what its economy would look like without this export market, which was worth $44 billion in 2019.[14]

Economist Ross Garnaut argues that we could not only fill that hole, but build huge future prosperity, by acting now on hydrogen. In his book *Superpower*, he posits that Australia has a unique opportunity to become a major global minerals-processing hub. We produce 53% of the world's iron ore, and with our abundance of sunlight, we could become the world's largest and cleanest steelmaker.

Australia also produces 40% of the world's bauxite (aluminium ore), and with clean electricity to process it, we could become the world's premier producer of clean aluminium. We also have large reserves of lead, lithium, manganese, gold, copper and other minerals. With clean energy to turn all of that ore into processed metal, the nation's future prosperity seems endless. But, if we are to grasp that opportunity before our competitors take an unbeatable lead, we need a visionary government pursuing long-term and consistent policy.

The first and very necessary step in realising this future prosperity is cleaning up the nation's electricity supply. But the fossil fuel industry is adamantly opposed to this, in part because it will lose market share, but also because doing so will open opportunities for products derived from hydrogen to compete with fossil fuels at every level, and in every market.

Hydrogen will keep popping up in this story, in ground transport, e-fuels and shipping. But the hydrogen story is even broader than that, for hydrogen can replace gas as a source of heat, fertiliser production and indeed almost every other product currently derived from fossil fuels. Clean hydrogen will be a vital tool in the many battles we need to win in the climate wars.

CHAPTER 12

Electric Vehicles

ONE of the most important lessons from the COVID-19 pandemic is the demonstration of how flexible even long-standing behaviour can be. Within weeks of the pandemic's appearance, individuals dramatically cut the amount of travel they did. Cars remained in garages, aircraft lingered in hangars and cruise ships went to home ports, while meetings were conducted via Zoom or similar means by people working from home.

There are signs that at least some of the changes the pandemic brought will be enduring. Working from home appears to be popular with many, and cycling and walking have seen an uptick. Zoom meetings may replace aircraft to some extent, and it's hard to imagine the cruise ships returning before a COVID-19 vaccine is developed. The most heartening aspect of this shift is the demonstration that by altering our behaviour several aims can be achieved simultaneously, bringing health and wellbeing benefits to individuals at the same time that fossil fuel use is reduced.

The impact of these behavioural changes on the oil industry has been substantial, forcing the price of oil to fall to economically unsustainable levels. Small oil producers started to go bankrupt, and large ones started abandoning projects. Indeed some analysts speculate that we may have already passed through 'peak oil'. But the pandemic has had potentially negative environmental effects as well, for it has also seen a flight from public transport and a drop in new car sales (including electric vehicles). The long-term impact on these sectors remains unclear, though some cities are promoting and planning for more walking and cycling as attractive alternatives to public transport and private vehicles, because in the COVID-19 era, both public transport and an increase in private vehicles on roads are unacceptable. This comes as electrification is revolutionising transport possibilities.

Buses powered by fossil fuels are noisy, polluting and expensive to run. Yet they were proliferating on our roads as, prior to COVID-19, demand for public transport increased. Buses are among the easiest of vehicles to electrify, as they travel on set routes over known distances, so charging can easily be catered for.

Right now, electric buses are cost competitive over their lifetime against buses powered by other fuel types. There is no need for a single new fossil-fuelled bus on Australian roads—we could quickly introduce e-buses on new routes and as replacements for old stock. But awareness of this opportunity is lacking, as is demand from the public. If you want faster action, contact your state transport minister.

Recently, a new and very exciting electric transport option has arisen. Trackless trams carry passenger loads similar to those of light rail, but don't need the fixed infrastructure such as tracks and overhead lines. They run at up to seventy kilometres per hour on rubber tyres, on conventional roads or dedicated paths. They are guided by laser, allowing for millimetre accuracy, and they carry their own batteries that can be recharged in thirty seconds while travelling, or in ten minutes at the end of their journey. Their bogies give an exceptionally smooth ride, and they are capable of carrying between 300 and 500 people. First developed in Europe and China, they're extremely economical, with estimated costs of between a third and a tenth of that of recent Australian light-rail projects.[1] Trackless trams are running on routes right now in Europe and China.

The COVID-19 pandemic has necessitated limiting the numbers of passengers in each vehicle, so the high capacity of trackless trams or light rail is useful. Moreover, we must plan for the future, and if we are to survive the climate emergency, that future must be electric and involve public transport.

The electrification of cars, and light and heavy trucks, seems to have been a long time coming. But it's worth remembering that Tesla, a market leader, was only founded in 2003, and it did not get its first electric car onto the road until 2008. Sales of electric vehicles (EVs) have increased rapidly—by 63% worldwide in 2017. Norway is the world's leader, with 60% of new car sales electric vehicles, and another 17% hybrids.

By 2018 there were about five million EVs on the road globally, with 45% of these in China. And 2020 looks to be the year of the EV, with a plethora of new models by many manufacturers becoming available. In Europe alone, the number of models will jump just from 100 to 175 during the year.[2] And costs are already coming down. By some estimates, EVs are already cost competitive with conventional cars in some markets.

In March 2020 ClimateWorks Australia, a leading climate-analysis group, stated that in ten years, at least half of all new cars on Australian roads would have to be electric, in order for Australia to achieve net zero emissions by 2050. Yet in 2019 full electric and hybrid car sales in Australia made up just 0.6% of total car sales in Australia, and full electric vehicles just 0.2%.[3] That's less than 100th the rate of the global leader, Norway. Even among the main countries that we trade with the uptake of EVs is 2%—ten times greater than the uptake in Australia.

A lack of proactive policy—indeed active opposition to electric vehicles by some Australian governments—means that Australia has lost seven precious years in the race to electrify its vehicle fleet.[4] Anyone who has travelled in Europe, Canada or the US will have seen for themselves how far behind we are.

There's a lot of misinformation propagated by the fossil fuel industry about electric vehicles, such as that the production of their batteries involves such massive greenhouse gas emissions that they pollute more than petrol-driven vehicles. This is simply untrue. The fact is that even if the electricity EVs run on is produced by burning coal, they'll pollute less than petrol vehicles. In fact, EVs are cleaner than petrol- and diesel-fuelled vehicles from cradle to grave.[5] And the cleaner their electricity supply, the less polluting their overall impact.

So what needs to be done if Australia is to make the EV transition in time to save the planet from 2°C or more of warming? In Australia, vehicle purchase price remains an important barrier. It's anticipated that by 2025 in Australia EVs will be price competitive with conventional vehicles, though a paucity of models available in Australia may mean that we take a few years longer to reach that point. But currently EVs cost about $15,000 more than comparable petrol models. It's worth noting that no nation has achieved high penetration of EVs without incentives that reduce the overall cost of purchasing and using EVs. In Norway, for example, EVs are exempt from a variety of taxes, tolls and parking fees.

One powerful incentive for EV uptake are the EU rules on vehicle emissions that came into force on 1 January 2020. They impose a limit on the average CO_2 emissions across all models offered by manufacturers. The penalty for exceeding the target is €95.00 per gram of CO_2. Were the penalty applied to 2018 sales figures, the penalty to car makers would have been £28.6 billion.

In Australia, an exemption from stamp duty (already in place in the ACT), would save $2000 on the price of a new EV. Matched with an up-front incentive of $5000, reducing by $750.00 each year, and the up-front cost disadvantage of buying an EV would be halved. Some of this cost could be clawed back by government with a road usage tax.

It would help uptake if a greater choice of models were available to Australians. In the UK, twenty models of EVs retail for between A$35,000 and A$50,000, but in Australia there are only four models in this price range. Measures that could see more choice for consumers include the Australian Government offering policy certainty, or fuel efficiency standards. For example, in Europe and many US states, car makers earn a credit worth A$3000–$6000 for each EV sold.

The other part of the solution concerns vehicle charging. Chargers are available across most of Australia, and many owners of EVs have them in their garages. But as the number of EVs grows, more will be needed. It costs about $1 million to install a fast charger that can provide 200 kilometres of range to an EV in eight minutes. If private investors contributed half of this cost, electrifying the entire national highway network with fast chargers would costs governments only $100 million.[6] In addition, there's a need to build urban charge points to cater for apartment dwellers who often don't have a way to charge their cars.

Because fleet purchases make up 50% of Australian new car purchases,[7] fleets are an important way of potentially increasing EV penetration. But it's virtually impossible at present. This is partly due to the way that the fringe benefits tax (FBT) works. Utes, for example, can be exempt from FBT, but EVs are not. And EVs don't get deductions for fuels and maintenance, because they don't use conventional fuel and require so little maintenance. In the grand scheme of things, the cost of fixing these problems and hastening the EV transition in Australia is very low. It could be paid for, with plenty left over, by simply by fixing anomalies in the current luxury car tax, which would result in A$3 billion of savings.[8] Specific policy adjustments include:

1. A detailed national EV study, providing details of electricity/engineering work.
2. A Productivity Commission review of vehicle taxes
3. Allow home solar/storage to be an FBT deduction
4. Reinstate the 'fuel efficient' gap under the luxury car tax
5. Develop EV targets for Commonwealth fleets
6. Establish a Federal Office of Electric Vehicles.

It's a great credit to the Government of New South Wales that it announced a major policy on EVs on 16 March 2020. The policy offers co-funding of fast electric-vehicle-charging facilities and enhances the ability of fleets, car-share companies and councils to purchase EVs.[9] As a result, right now, New South Wales is leading the nation in encouraging EV uptake and offering a much needed ray of hope in combating climate change.

CHAPTER 13

Aviation and Shipping

ONE of the astonishing paradoxes of COVID-19 is that it has had a massive impact on emissions from the very sectors that experts have long considered to be the most intractable. Between 1 February and 19 March, CO_2 emissions from aviation were 10 million metric tonnes lower than the same period in 2019.[1] This is a huge reduction. But when combined with the total cessation of the cruise ship industry, COVID-19's impact on transport emissions becomes gargantuan. It's worth remembering that this drop in fossil fuel use was caused by an abrupt behavioural change. It should remind us of how quickly we can address the climate emergency simply by altering our behaviour.

It's likely that in the long term, both aviation and shipping will recover, so if we are to keep the emissions gains, it's vital that aviation and shipping return clean and green. Because of its reliance on high-density energy sources, aviation will probably remain dependent on fossil fuels the longest of all forms of transport. But even here there are signs of hope. The first commercial electric-plane test flight occurred in Canada in early 2020, when a De Havilland seaplane retrofitted with an electric engine and batteries took to the air over the Fraser River in British Columbia.[2]

One single test flight of a light aircraft is a long way from reducing fossil fuel dependence in aviation. But there's interest in the prospect of future electric flight, with about 170 electric-aircraft projects underway. And the interest in emissions-free travel comes not a moment too soon: over the past five years, emissions from airline travel have risen 23%.[3]

It's not clear how long the COVID-19 pandemic will cast a shadow over the aviation industry. But companies could use it as a valuable opportunity to review their options to reduce their environmental impact. A recent survey by Griffith University's Dr Susanne Becken of fifty-eight major airlines found that none is doing enough to reduce emissions.[4] While many are increasing carbon efficiency (carbon burned per seat), none is reducing emissions overall, and the problem is large. Qantas alone is responsible for 12 million tonnes of CO_2 emissions annually. The study identifies twenty-two opportunities for airlines to cut emissions, and all should be pursued. Tellingly, electric aircraft are not listed among the opportunities.

While some airlines are pinning their hopes on electric aircraft becoming commercially available, as yet only one model is under development—the Airbus E-Fan X hybrid—and it's still at a very early stage. One engine will be electric on its maiden flight in 2021, with the other three powered by conventional jet fuel.

There's a good reason for the slow development of electric planes. Jet fuel is power dense—around 12,000 watt hours per kilogram—while batteries can provide barely a sixth that. For small aircraft flying short distances, the difference isn't so critical. But for large, heavy intercontinental planes, the low energy density of current batteries makes them unviable. As a result, nobody expects to be flying long distance in electric aircraft before mid-century, if ever. And that is just far too late in terms of the climate crisis. The climate emergency has entered a critical stage, and to delay dealing with aviation emissions until electric jetliners are available may mean that we lose the battle to stabilise the climate.

Another way must be found. Aviation is one area where hydrogen is unlikely to be the answer, simply because it's not sufficiently energy-dense, though hydrogen might be used as a component in future jet fuels. Another possible solution lies in biofuels. During World War II, as the Germans ran short on oil, the Luftwaffe reputedly ran in part on the fuel butenol, which was refined from potatoes, so the idea of using biofuels in aircraft is not new. Now, some companies are looking at what contribution biofuels may make to powering contemporary aircraft. In April 2019 Scandinavian Airlines (SAS) and Swedavia (the company that runs Swedish airports) announced that all domestic flights in Sweden should be running exclusively on biofuels by 2030.[5] By 2045, international flights from Sweden should also be fuelled by biofuels. But as SAS's CEO Rickard Gustafson says, 'Today's aircraft can fly on biofuel. The problem is that it is not possible to buy enough of it.'[6] And building volume is a formidable barrier. To achieve their targets, the Swedes would need to source 11,000 tonnes of biofuel in 2021, rising to 340,000 tonnes in 2030.[7]

Sweden is looking at waste from its large forestry industry to provide the required volume. This waste has been found to be suitable for conversion into biofuels, and with the addition of other biological waste, the Swedes are confident that they can develop a biofuel industry of a sufficient scale to satisfy their future domestic aviation needs. 'If Sweden takes the global lead regarding biofuel production, we can create a brand-new industry and many new jobs in Sweden,' says Swedavia CEO Jonas Abrahamsson.[8]

It might seem, from a climate perspective, that biofuels have little advantage over jet fuel made from oil, because the burning of both creates CO_2. But in fact biofuels represent a huge leap forward, because the carbon in them comes from the biosphere (trees), which means the perpetual cycling of carbon between trees (which draw carbon from the atmosphere to grow) and the atmosphere, rather than fossil deposits, the carbon in which adds to the net amount of carbon in the biosphere when it is burned.

Although biofuels may prove hugely important as a bridge, particularly in domestic aviation, at the global scale they are unlikely to be the answer. Ultimately, the best solution for aviation may lie in the development of e-fuels. E-fuels are jet fuels made from atmospheric CO_2 and hydrogen that have been generated by using clean energy. One of the frontrunners in the manufacture of e-fuels is the Canadian company Carbon Engineering. Carbon Engineering's e-fuel is made by sucking CO_2 out of the atmosphere and combining it with hydrogen derived from water using British Columbia's abundant hydropower to produce energy-dense e-fuels.

Carbon Engineering's e-fuel has a high cetane rating (meaning it burns well in engines), it can be blended with other fuels and it doesn't contain pollutants such as sulphur, nitrogen or particulates. It costs less than US$4 per gallon to produce this e-fuel, which makes it more expensive than Jet Fuel-A (a fossil fuel), but about the same as biofuels.[9]

The biggest roadblock to e-fuel of the type Carbon Engineering manufactures, is access to sufficient cheap, clean energy to produce the hydrogen required. Australia has stupendous resources of solar and wind, so we are well positioned to break through that barrier. What is needed now is a plan, similar to Sweden's biofuel initiative, to make e-fuels a reality for Australian aviation.

Qantas could help in achieving this, for the airline has good carbon reporting, being one of only eight airlines to use the 'Task Force on Climate-related Financial Disclosures' process. Developing a plan to use e-fuels on domestic flights, on time scales similar to those of Swedish airlines, would be a massive first step. As the national carrier, Qantas also has a strong voice. Working with the Federal Government and airport owners, it could devise, and commit to, a plan to be carbon free by 2050.

For a year I took the Manly Ferry to work. As I crossed Sydney Harbour, I was very aware of the large volumes of diesel fuel being burned daily by the city's ferry fleet. I've been to Hong Kong and enjoyed travelling on the electric ferries there. Why can't Sydney follow Hong Kong's lead, and start decarbonising its iconic ferries?

Approaching Circular Quay, another maritime blackspot came into view. Until the arrival of COVID-19, there always seemed to be a cruise ship in port, belching some of the worst levels of pollution humans create by burning fossil fuels into the heart of the city. Until COVID-19, the cruise-ship industry was booming, and cruise ships, along with the 50,000-odd merchant ships plying the oceans, almost all run on bunker fuel—the dregs of the oil refinery process and one of the most polluting of all fuels. As the cruise-ship industry restructures, it needs to clean itself up. From poor regulation under 'flags of convenience', to its employment practices, there's a lot for the industry to do. But one thing that should be on the agenda is emissions while in port. For the emissions from moored cruise ships are breathed in by millions.

Since 2015 ships mooring in Sydney Harbour have had to use low-sulphur bunker fuel. But this fuel is still polluting and we can do far better. A large number of options are available, and one of the best is the provision of electricity to the moored vessel so that it can plug in and turn off its bunker-fuelled generators.[10] A cruise ship facility at Circular Quay would require the upgrading of a nearby electricity substation and provision of other infrastructure. And of course the cruise ships would need to be fitted so that they could plug in.

It would have little impact on emissions for just a single port to do this. But a national program could make a difference if any ship visiting anywhere in Australia were required to comply. There is plenty of precedence for this approach. California has demanded that shipping use shore power or equivalent measures since 2007, and the practice is common in other places in North America, Europe and elsewhere.[11] Again, Australia is lagging.

Ninety per cent of international trade relies on shipping, so as global trade grows, so do emissions from shipping. Shipping currently accounts for between 2% and 3% of global CO_2 emissions. Prior to the COVID-19 pandemic, global trade was expected to grow 4% per year to 2023.[12] That increase is almost certain to occur at some stage. And with the International Maritime Organisation's plans to reduce emissions from shipping by at least 50% by 2050 through the introduction of zero-emissions vessels, the time for action is now.

Financial institutions are playing a role in encouraging this transition, through the Poseidon Principles, a framework financial institutions can use to assess the carbon footprint of vessels as new financing is required. The Poseidon Principles create an incentive to finance greener vessels. It's important to ask your financial institutions to support the Poseidon Principles. But that is only part of the solution.

Biofuels are unlikely to be used to power shipping, because of both the limited land area available to grow biofuels at the volume required and the transport costs involved. Ships aren't constrained by low-energy-density fuels in the same way aircraft are, so they can use hydrogen as a fuel source directly, without having to convert it into e-fuels. The best solution for future ship fuel currently appears to be hydrogen.

Converting shipping to run on hydrogen fuel in the timeframes required to address the climate emergency will require rapid development of the clean hydrogen economy, and the financing of port infrastructure to store, distribute and possibly produce hydrogen. Financing of these initiatives could be facilitated by state governments and port authorities. But because shipping is a global business, with ships needing to be refuelled in various international ports, an intergovernmental agreement could expedite the shift.

This all seems rather distant, but the basic technologies required to move from bunker fuel to hydrogen exist today. Indeed, a new breed of hydrogen-powered vessels is on the horizon. In February 2020 it was erroneously reported in the media that Bill Gates had purchased the world's first hydrogen-powered super yacht. The $1 billion vessel, which is under development by the Dutch firm Sinot, is a real project, even if Gates's supposed purchase was not. Sinot is promoting the concept through events such as the Monaco Yacht Show.

Plans are also afoot for the first hydrogen-powered cruise ship. The ship, owned by the Havila group, currently tours Norway's fjords, powered by conventional shipping fuels. The retro-fit for hydrogen fuel will be complete by 2023.[13]

As hydrogen replaces gas as an energy source, it will have to be bulk shipped, just as gas is today. Hydrogen could be transported in the form of ammonia, but a more practicable option is liquified hydrogen, and this will require the construction of an entire fleet of hydrogen bulk carriers. Although it doesn't run on hydrogen itself, the world's first liquified hydrogen carrier, the *Susio Frontier*, was launched in 2019. It will be used to carry hydrogen from Australia to Japan, but, tragically, the hydrogen it transports will be produced from brown coal.[14]

As we contemplate the enormous investments required to make global hydrogen trade a reality, it's worth remembering that just a few decades ago a global trade in gas barely existed.

The first liquid-natural-gas-transport freighter, the *Methane Pioneer*, was launched only in 1959, and the large-scale shipping of gas dates back only to the late 1970s.

CHAPTER 14

Leadership and Cooperation in a Global Emergency

THE COVID-19 pandemic has taught us a lot about cooperating in an emergency. Each nation has dealt with the arrival of the virus in its own way. But nations have quickly learned from each other about what works and what doesn't. Except for the advisory role played by the World Health Organization, there is no globally-directed response to COVID-19. But nevertheless most nations are making progress at containing the disease.

Of course, the pandemic has also seen a rise in friction. Nationalist leaders are always looking for others to blame and excuses to close borders. But even they cannot deny the imperative for global cooperation when it comes to the development of a vaccine, for the task is so huge and expensive that timely vaccine development is beyond the reach of countries acting alone.

There is a remarkable parallel in all of this with the climate emergency. It's true that a global agreement—the Paris Agreement—already exists to combat climate change, but it's a voluntary, bottom-up approach with no sanctions for those who underperform. Nations are doing things in their own way, but learning from each other as the climate emergency unfolds. And, as with COVID-19, China is playing a central role, for it is the world's manufactory for key elements of the energy transition, including solar panels, batteries and electric vehicles.

Australia has a proud history of global leadership on environmental issues, going back to the first Antarctic Treaty in 1959. It has played a key role in protecting Antarctic flora and fauna, regulating fishing, aviation and shipping in Antarctica, outlawing pollution, and in keeping the Antarctic nuclear free. Australia was given the honour of hosting the first Antarctic Treaty consultative meeting in 1961. Two decades later, Australia played a critical role in the development of the Convention on the Conservation of Antarctic Marine Living Resources. This offshoot of the Antarctic Treaty, which came into being in 1982, is based in Hobart and continues to play a crucial role, particularly in regulating the harvest of krill.

Few challenges are as diplomatically difficult as regulating activities across an entire continent over which seven governments have asserted multiple, overlapping sovereignty claims. Australia's long and proud history illustrates just how respected, competent and effective Australian governments and negotiators can be in the international arena. If that determination and ingenuity could be applied to the climate problem, Australia could achieve a great deal in bringing the governments of many nations together to avert climate catastrophe.

Australian governments and businesses could begin by leading in the many international fora that already exist. For example, the International Maritime Organisation (IMO) for shipping, the International Air Transport Association (IATA) for air travel, and the International Association for Hydrogen Energy are all important organisations where Australia could lift ambition and sharpen focus. But the key forum for reducing emissions is the COP (Council of Parties meetings, under the UNFCCC). And there Australia's performance over the past eight years has been worse than abysmal.

Ever since the election of the Abbott government in 2013, Australia has frustrated attempts at emissions reduction through the COP negotiating process. In doing so it has frequently acted in concert with the US, Saudi Arabia and other nations that benefit financially from fossil fuel exports. This fossil fuel clique has proved highly effective in stymying progress. Whenever I've attended recent meetings of the COP, the antics of this bloc has reminded me of the promoters of the nineteenth-century opium trade. And currently, Australia's reputation among other nations is about as low as that of a drug dealer.

The anger of those interested in addressing the climate emergency is expressed in many ways, but one of the most telling is the announcement of the 'Fossil of the Day Award'. Given by the Climate Action Network, it is symbolic of disgust felt by many at those aggressively seeking to extend the use of fossil fuels. During the 2019 COP meeting, Australia received a Fossil of the Day Award for 'using carbon market loopholes to meet climate targets', a reference to Australia's attempt to claim 'carryover' credits from an earlier treaty, the Kyoto Protocol, and for threatening to withhold funds owing under the Green Climate Fund.[1]

Sadly, it is not just Australia that has failed, but the COP process itself. The COP negotiations progress through annual meetings, known as a COPs (starting with COP 1 in Berlin in 1995), hosted by a different country each year. The COP meetings have now been running for a quarter of a century, and so very little has been achieved at meetings dealing with what most agree is a crisis. The length of the negotiations is itself an emblem of our collective failure to date.

But we could do things so much better. Christiana Figueres, the UN executive secretary for climate change and the public face of the successful 2015 Paris Agreement, knows the problem of the COPs intimately. As she puts it, for the past twenty-five years the COP negotiations have been framed as 'I win, you lose' arguments.[2] While clean energy sources were more expensive than polluting ones this was inevitable, because anyone investing in clean energy had to bear the extra economic cost.

The 'I win, you lose' mentality has ramified into many narratives that waste an enormous amount of time at the COP meetings. One of the most frustrating has been the concept of 'historic responsibility', which has consumed more words and hours than any other I remember. The doctrine of historic responsibility posits that the industrialised nations are, in historic terms, responsible for most of the greenhouse pollution in the atmosphere, and that they therefore need to take on the greater part of the burden in reducing emissions, as well as compensate the least developed nations, who are feeling early climate impacts.

While this doctrine may have been useful in 1995, it has not been reflective of the reality for some years. Half of the greenhouse gases ever emitted by humanity have been emitted during the last thirty years, and the largest polluters today are China and India, neither of which falls under the banner 'industrialised nation'. So, the doctrine can easily become an excuse for those countries to continue to pollute until their standards of living equal those of Europe and the USA.

The argument also posits that developed nations should be assisting less-developed nations as they deal with the consequences of climate change. This is a very real need, because some of the least-developed nations face overwhelming, climate-related problems and have little means to address them. But although this is an issue that needs serious attention, it should not hold up global negotiations on reducing emissions.

Fortunately, new developments in the economics of clean energy have the potential to bypass these entrenched battles, by separating the deployment of clean energy from arguments around historic responsibility and current and future compensation. These same developments have emphatically destroyed the old paradigm of 'I win, you lose' that has so long bedevilled the COP meetings.

Steep and ongoing reductions in the costs of clean energy, in almost all countries, mean that new wind and solar power has become less expensive than new fossil fuel power systems. And in many countries it's already cheaper to build new wind and solar than to keep old coal plants running. We have moved from an 'I win, you lose' world, to a 'win–win' one.

But because of the new 'win–win' world created by cheap renewables, the potential exists for influential nations to introduce something new into the COP process. And arguably no nation is better positioned to achieve this than Australia. Change will not be easy. But as Christiana Figueres sees it, 'The most powerful thing you can do is to change how you behave in that landscape, using yourself as a catalyst for overall change.'[3]

Were Australia to follow Figueres' advice, we could carry the entire Pacific with us, as well as many other nations that seek earnestly to avert the climate emergency. As I think back on

Herbert Vere (Doc) Evatt, an Australian and the fourth president of the UN General Assembly (1948–49), who co-drafted of the Universal Declaration of Human Rights, I can see how a young nation with a new narrative can shake things up.

The far right of the Australian Coalition Government is suffering from its own virulent infection—a denial in the face of all evidence of the threat of climate change. This ideology is increasingly unpopular and unsustainable. The COVID-19 pandemic has shown us that the Morrison government can follow scientific advice, and do what is normally unthinkable, when faced with a crisis. The climate emergency is slower burning than COVID-19, and doesn't often threaten individual health as broadly or immediately as the virus, but it is at least as deadly, and delaying action will have dire consequences. Whether the denialist infection of the far right can be vanquished by the experience of COVID-19 remains to be seen.

But if it can be, Australia could change its negotiation position. It could also tell an extraordinarily powerful 'win–win' story, for its wind and solar resources are among the best in the world, and innovations such as South Australia's big battery (of which there are now three) are deeply inspiring. And were we to grow the hydrogen economy, we could be the change the world needs to create if it is to move on to greater wealth, health and climate security.

CHAPTER 15

Adaptation

THE climate emergency is now so dire that future COP meetings must focus on far more than reducing emissions. To avoid catastrophic damage, the world needs to develop the equivalent of 'emergency rooms' (ERs) to shelter the worst effected of Earth's ecosystems, as well as to assist the worst affected human populations. Developing both ecosystem and societal ERs will be hugely expensive.

One attempt at setting a societal ER is the Global Climate Fund (GCF), which was established at COP 16 in Mexico, in 2010. It aims to raise US$100 billion by 2020 to invest in helping countries adapt to and mitigate climate change. Australia stopped making payments to the Global Climate Fund (GCF) in 2019, and with hostility to the initiative from the Trump presidency, it's unlikely that the GCF will achieve its goal. Other ways must be found to protect the most vulnerable.

The provision in Zali Steggall's climate bill for a national risk assessment in the face of climate change could be the beginning of a national climate emergency response, for it would inform us of which people and assets, and where, are most at risk.

Arguably the most important asset in keeping us safe is the knowledge of which threats are likely to materialise and when, so that we can prepare for them by developing detailed, science-based strategies. As adaptation experts Tayanah O'Donnell and Josephine Mummery put it, 'Lives and money will be saved by strong climate adaptation measures.'[1]

When it comes to adaptation, Australia is failing not only the poorest and most vulnerable nations, but itself as well. This was not always the case. Back in 2007, Australia made a strong start on climate adaptation, and it is one of the nation's great tragedies that our efforts foundered amid ideological opposition.

Under the governments of John Howard, Kevin Rudd and Julia Gillard, a robust institutional framework was put in place to deal with adaptation. Key to it all was the CSIRO's Climate Adaptation Flagship. Tasked with equipping Australia 'with practical and effective adaptation options to climate change and variability and in doing so creating $3 billion per annum in net benefits by 2030', it had 160 full-time staff and an annual budget of $43 million.[2] The flagship took a holistic approach, looking at adaptation in industries, cities, biodiversity and primary production.

In 2010 a research arm was created by the Federal Government. The National Climate Change Adaptation Research Facility (NCCARF), with a budget of $50 million, provided Australia with a powerful engine to guide investment and policy. Tragically, both the flagship and NCCARF were emasculated by the Abbott government: the CSIRO flagship was 'restructured' in 2014, and funding for NCCARF cut to $9 million. Then, in the 2017 federal budget, funding was withdrawn entirely.

Subsequently, as emissions have accumulated in the atmosphere and the climate has deteriorated, Australia has been flying almost blind as far as adaptation is concerned. As I write, all federal climate-related policies remain under review, including adaptation. The one partial exception to Australia's destruction of adaptation measures seems to be in primary production, where the Federal Government maintains a focus on assisting farmers adapt to climate change.[3] Overall, however, Australians are far more poorly equipped to adapt to unavoidable climatic change today than we were a decade ago.

Despite the catastrophic and wilful failure of federal leadership, Australian farmers, Indigenous communities and many others have been adapting to our changing climate for decades. But the 2019–20 megafires illustrate how vital it is that government plays its role. If we had been prepared with a strategy informed by experts, much damage may have been avoided.

The climate challenges we now face are so great that we must resurrect the coherent, holistic approach destroyed by the Abbott government. For no matter how successful we are at reducing greenhouse gas emissions, and indeed at drawing carbon down from the atmosphere, the task of adapting to inevitable climatic change remains enormous.

NCCARF should be reinstated as a matter of urgency, so that decision-making on funding and research directions has a scientific base. It is important as well that adaptation to inevitable climate change seeps into the policy-making of all government departments. The issue should be seen very broadly, from adapting our institutions (from finance to emergency services) so that they better serve us in the changed climate, to adapting our infrastructure (which may include the abandonment of some of it), and the adaptation of our human populations.

When it comes to expenditure, national adaptation policy should be used as a sieve. High priority areas should receive preferential funding, while expenditure on infrastructure in locations that will be threatened by climate change, and which are likely to have to be abandoned, for example, would not pass the sieve test. Likewise, expenditure on infrastructure that is unable to function properly in the future climate should not be funded.

The area is complex, regionally specific and detailed. In

general terms the following is what is required in some of the most urgent and impactful areas.

For years firefighters have known that fire seasons are lengthening; the number of very severe fire-danger days is increasing, and the opportunities to undertake controlled burns are decreasing. In the face of this spiralling risk, fire chiefs called on the Federal Government to investigate whether the nation's firefighting services were adequately resourced, particularly with regard to expensive items such as firefighting aircraft, which are generally too costly for individual states to purchase and maintain. Prior to the catastrophic megafires, these pleas fell on deaf ears.

Fire chiefs have also called on state and territory governments to cease cutting budgets for foresters, parks rangers and volunteer firefighters, and instead to increase resources to these groups to enable them to undertake fuel-reduction activities in the diminishing time windows now available to them.[4]

The catastrophic megafires of 2019–20 revealed how woefully underprepared the nation is in its ability to respond to the new fire reality in Australia. Had we acted earlier, lives, property, forests and money could have been saved. Given the inevitability of megafires in future, Australia urgently needs a nationally coordinated approach to the acquisition and use of expensive firefighting equipment such as aircraft, and much better funding and co-ordination of fire services at the state level.

Furthermore, the government's response to communities,

industries and individuals affected by fires is inadequate. Given the inevitability of future severe fires, the Federal Government should create, with the support of state governments and local councils, a fit-for-purpose entity that optimises the chance of swift recovery for fire-devastated regions.

Already, experts are warning that heatwaves threaten to overwhelm the world's hospital systems.[5] Representatives of the emergency, health and urban services sectors warn that Australia is underprepared for future heatwaves.[6] If Australia does nothing, our health services will be unable to cope with the victims of future climate catastrophes such as heatwaves. This is because individual heatwaves are getting longer, their seasons are extending, and the peak temperatures are increasing.

Already, heatwaves kill more Australians than any other natural cause. Liz Hanna, President of the World Federation of Public Health Associations, says that more Australians now die of heat stress than die on the roads. Australia has aggressive policies to stem the road toll but, Hanna says, as 'heat-related deaths in Australia climbed into the thousands every year on average, but little was being done about it'.[7] As a nation, we must prepare an equally ambitious adaptation strategy for stemming the escalating climate death toll caused by heatwaves.

Among the priority areas for action is increasing public awareness of the growing danger of heatwaves. A campaign, based on the road-toll campaign, could go a long way towards preparing the community for the danger. But there is a risk that the denialists in the Federal Government would block this.

With a warmer atmosphere and greater rainfall intensity, floods and cyclones are also increasing in intensity in Australia. And the drying trend across southern Australia is revealing deficiencies in our water supply systems. Some, particularly climate denialists, promote more dams as the answer to Australia's climate-caused water problems. In fact, big dams create more problems than they solve. A study by the International Rivers Organisation has shown that as rainfall becomes more variable in our changing climate, big dams turn from being assets to massive liabilities. That's because they cannot afford to release water for hydro-power generation to the same extent as when rainfall was more predictable, they have shortened lives due to increased siltation, and they have become increasingly at risk from collapse due to catastrophic flooding.[8]

In the face of these multiple assaults on our society, I believe that the best defence is the establishment of a properly constituted, apolitical National Commission for Climate Adaptation. The first step would be to undertake, as Zali Steggall proposes, a national climate risk assessment. The commission should have the authority to direct research by a resurrected NCCARF, to estimate the damages being inflicted on Australia by climate change, and the authority to prioritise funding and actions. Such an entity is unprecedented in Australia. But the COVID-19 pandemic has seen much unprecedented action.

Perhaps the greatest and potentially most costly threat in the long term stems from rising seas. But because the rise is to some extent predictable, we are well-placed to deal with this threat in a timely manner. Australians are also fortunate that a tool for forecasting coastal erosion in the short term has recently been developed. The tool calculates imminent risk to coastal communities in ways that assist in allocating relief funds in a timely manner.[9]

Some damage caused by rising seas is inevitable, and abandonment may be the most cost-effective or only viable action in some circumstances. It is also critically important that we understand that we can control, to some extent, how fast the oceans rise. The greater our greenhouse-gas emissions, the faster the rise will be. By reducing emissions, we will be buying ourselves precious decades and centuries to deal with the problem.

The large-scale adaptation required by sea-level rise will be expensive. Other nations faced with inevitable damages from natural disasters have provided for them beforehand. New Zealand, for example, has a Natural Disaster Fund (NDF), which levies $0.20 for every $100.00 worth of insurance cover taken out in the country. Australia could develop a Coastal Defence Fund, along the lines of New Zealand's NDF, to be administered by experts and used for restoring or relocating infrastructure at risk of sea-level rise. Some may argue that, given the escalating cost of insurance, a Coastal Defence Fund would simply be too expensive. But Australian governments waste huge amounts of money dealing ineffectively with manifestations of the climate crisis. And without a comprehensive plan, they will continue to do so.

Because vulnerability of infrastructure to sea-level rise is to some extent predictable, the fund could be expended judiciously, in a 'just in time' manner. The fund could even assist where a planned abandonment of indefensible assets is part of the national strategy. At present, if your land is claimed by the sea, you've lost it. A Coastal Defence Fund could be used to offer some recompense.

In its recent bushfire recovery package, the Federal Government provided funds to restore biodiversity affected by the fires. Given the multiple challenges biodiversity faces, the nation will require similar funding on a long-term basis.

But more lateral thinking will be required of our biodiversity managers as well. Climate change will leave some species without viable habitats. We will then have no option but to take them into captivity, or to relocate them to areas of suitable habitat. At the moment, many biodiversity managers understand this, but are reluctant to move forward aggressively with translocations.

Finally, Australia needs to adapt its legal systems to the new climate to provide a proper legal framework that allows those suffering damage from the changed climate to sue those who are accessory to inflicting it on them. It might also cause those who are damaging our future, and those of our children, to think twice before continuing to pollute.

Such changes are already nascent in Australia. In April 2020 a group of bushfire survivors issued a legal challenge to the New South Wales Environment Protection Authority, seeking to force it to act on climate change in order to protect communities from catastrophic fires.[10] This is the first action of its kind in Australia, but it is unlikely to be the last.

We cannot afford to forget the environment's casualties, for the climate emergency is far advanced, and parts of our planet are already in crisis. Where possible, we need to get these most imperilled environments on the equivalent of life support. In Australia, steps are already being taken to do this with what is arguably the nation's most endangered habitat—the Great Barrier Reef. We have, scientists warn, no more than a generation (thirty years) before the reef will be no more. Given the scale of what needs to be achieved, this is a brief time indeed. Again, we find ourselves acting at the last possible moment.

The Federal Government, via the not-for-profit Great Barrier Reef Foundation, is funding research into forty-three ideas that might offer a lifeline to the reef, including brightening clouds with salt crystals so they reflect more light and so cool the area below them; stabilising damaged reef structures using concrete or 3D-printed forms; seeding the reef with hand-raised coral larvae; and breeding corals that are heat-tolerant.[11]

Brightening clouds is a form of geoengineering that can be turned on and off. At best it will only partially alleviate the plight of the Great Barrier Reef, for the heat that affects it can role in from the Pacific in submarine heatwaves, as well as being generated by sunlight hitting the reef lagoon. And breeding heat-resistant strains of coral may involve genetic engineering that would permanently alter the nature of the Great Barrier Reef. Such measures would, at other times, merit wide scientific and public debate. Yet given the extreme urgency we face, I feel torn. Would a doctor, facing an emergency, debate at length about what action to take, or act in what they see as the way most likely to provide relief?

We must also recognise that, just as placing a patient on a ventilator is an interim measure, most of the actions proposed to save the reef are only stopgaps until the crisis passes. And the crisis the Great Barrier Reef faces will only be passed when we develop the facility to draw CO_2 out of the air at volume, which will take decades to develop. Fail to do that, and we shall gain nothing for all the money spent on emergency measures.

At a global scale, some of the most endangered environments are those whose destruction could push Earth's climate system beyond the tipping point into a new, hostile state. The rapid pace at which these global tipping points are being approached has been highlighted in a recent study.[12] One critically important endangered environment is the cryosphere, as Earth's icy regions are known. If any habitat on earth deserves an ER response, it is our ice. At current rates of ice loss, tipping points in the cryosphere could be reached as soon as this decade. If the Arctic becomes ice free in summer, for example, melting of the permafrost could accelerate, as could melting of the Greenland ice sheet. This would herald the eventual loss of the world's coastal cities and alter the shapes of the continents themselves, as well as creating a positive feedback loop, where warming releases vast volumes of carbon and methane.

Creating an ER for the cryosphere would require a global effort, the politics of which would be difficult, to say the least. And again, we find ourselves needing to act almost at the last possible moment. How could, and should, humanity respond? Scientists have given considerable thought to how the cryosphere might be stabilised. Sergey Zimov, a Russian geophysicist who specialises in arctic and subarctic ecology, established a zoo known as Pleistocene Park in Siberia. It is stocked with hoofed animals such as bison, horses and reindeer, and researchers have discovered that they have a significant cooling effect on the permafrost by trampling snow, which reduces the snow's insulation of the ground below from the freezing air above. Scientists believe that the warming impact of climate change on the permafrost might be halved, and 80% of the permafrost protected until at least 2100, if large mammals were re-introduced throughout the Arctic.[13]

Other researchers believe that a partial solution, at least, lies in geoengineering. Geoengineering involves the manipulation of aspects of the environment in an attempt to counteract the effects of global warming. The idea arguably originated with atmospheric chemist and Nobel laureate Paul Crutzen who, in 2006, called for research into injecting sulphur into the stratosphere to cool the planet.[14] Some scientists believe that Crutzen's idea offers an inexpensive, instantly effective measure.

Several national and multinational geoengineering research projects have already been established to investigate this idea, and a number of government institutions have commissioned assessment reports.[15] Cloud brightening, as suggested for the Great Barrier Reef, is a possibility. But the most favoured method involves mixing sulphur into the jet fuel of aircraft to be expelled with the exhaust as they fly over the Arctic. Jetliners fly in the lower stratosphere, so the sulphur would be injected directly into this part of the atmosphere, where it would reduce the amount of sunlight reaching the pole. This would cool the region, and so maintain the ice.

Here we face a profound problem. With the science of geoengineering advancing by the year, and the threat of sea-level rise growing, our capacity to act is now well in advance of both our politics and our understanding of the full consequences of action. I greatly fear that some nation may decide to unilaterally implement a poorly researched geoengineering solution. So much basic science surrounding various geoengineering options is still lacking that we cannot be sure such options won't create more problems than they solve. And because the consequences will affect the entire globe, the potential for conflict is large. The computer models currently used to assess the impact of stratospheric sulphur on Earth's climate system come from studies of volcanoes, particularly well-studied eruptions such as that of Mount Pinatubo in the Philippines in 1991, which released 15 million tonnes of sulphur into the stratosphere. In the fifteen months following the eruption, Earth cooled by 0.6°C.[16]

At present, there have been no studies on how the addition of sulphur to jet fuel over the Arctic might differ in its impact on the climate from the injection of sulphur by tropical volcano eruptions. There is an urgent need for such studies because volcanoes and jet aircraft exhaust streams are very different, and much is at stake. For example, changes in atmospheric circulation can affect the monsoons, and in south Asia alone 1.4 billion people are dependent on the life-bringing monsoon rains to grow food crops.

There is very little time for action. It may be that nations will follow the lead of some in Australia who seek to protect the Great Barrier Reef using geoengineering, and start altering clouds over the Arctic, or using sulphur to shield the ice from sunlight.

So what should Australia do to help prepare the world as we slide towards such emergency action? Australia's climate science and modelling capabilities are among the best in the world. We could even establish our own national research centre, tasked with studying a variety of geoengineering options so that the full consequences of geoengineering are understood before they are trialled.

One of the greatest obstacles to geoengineering our way out of our self-created climate problem is the lack of political agreement. Australia, along with like-minded nations, could act on the global stage as deal-brokers and mediators, helping convene groups of nations tasked with deliberating on rules, guidelines and protocols.

CHAPTER 16

Drawdown: A Vaccine for the Fossil Fuel Pandemic

DRAWDOWN, also known as atmospheric CO_2 removal, involves the removal of CO_2 from the atmosphere and storing the removed carbon in ways that prevent it from returning to the atmosphere for decades, centuries or longer. I have written extensively about the various drawdown options in my books *Atmosphere of Hope* and *Sunlight and Seaweed.*

Drawdown is what trees, seaweed and other photosynthesising organisms do. But some kinds of rocks, as they decompose, do the same thing. The impact can be large: in total, plants and rocks draw down around half of the CO_2 that we put into the atmosphere each year. Industrial process can also draw CO_2 from the air, as can certain kinds of concrete.

Many people—even some of those involved in climate science—regard the drawdown of CO_2 from the atmosphere to combat climate change as unrealistic, because the volumes of CO_2 that must be withdrawn are so great. Others oppose it because they believe that the fossil fuel industry will use the promise of drawdown as an excuse to keep on polluting, just as the proponents of hydrogen from coal have done in the Latrobe Valley.

Incidentally, drawdown pathways is a better term than 'drawdown technologies' as not all drawdown opportunities are technological. But all are pathways that begin with the CO_2 in the air and, through sometimes overlapping means, end with the carbon in the CO_2 locked safely out of the atmosphere.

Drawdown is important for our climate security because it can slow global warming by reducing the concentrations of greenhouse gases in the atmosphere. If done at sufficient scale, it may even be able to partially reverse that warming. But we are very far from that point today, for the scale of drawdown required to achieve this is truly enormous. We'd need to take eighteen gigatonnes of CO_2 out of the atmosphere to reduce atmospheric concentrations of the gas by one part per million. To do that by planting trees, we'd need to plant an area at least twice the size of Australia in vigorously growing forest.

Of course it's a great idea to plant trees and protect forests. Such actions have many benefits, from protecting biodiversity and water sources, to improving agriculture. But there are several reasons why trees are unlikely to provide a solution to the climate emergency. The first is the scale of the problem. Humans currently emit about fifty-five gigatonnes of CO_2 into the atmosphere every year. To take out just five gigatonnes per year by planting trees, we'd have to cover an area a little smaller than Australia in flourishing forest, and let it grow for a century.

Regardless of how strongly we act to reduce emissions, the next few decades will see ongoing deterioration of the global climate, and a growing threat to our existence. It is generally agreed by scientists that global temperatures are likely to rise to about 2°C above pre-industrial levels this century, even in the best-case scenario. So how can we stay below the widely agreed safe threshold of less than 1.5°C? The only possible option is to cut emissions drastically, and to simultaneously begin to draw gigatonnes of CO_2 out of the atmosphere.[1] And even then we will overshoot and have to reduce greenhouse gas concentrations quickly.

The issue of whether the fossil fuel industry will use drawdown as an excuse to keep polluting deserves serious consideration. There is a strong argument that, during the thirty years leading up to the failed Copenhagen climate meeting, the idea that carbon capture and geological sequestration (CCS) technologies will be developed in future has lessened the urgency of reducing greenhouse emissions.[2]

Today we live in a world where CCS as it was conceived of at that time is a clear failure. Despite the fact that the Federal

Government is showing renewed interest in this approach, it should not be pursued. But new forms of CCS offer real promise. The biggest change since 2009 is that the climate problem has grown so large that drawdown at a large scale is required to achieve climate security. Should we ignore drawdown in the full knowledge that it is essential to stabilise the climate? Or should we recognise that it's indispensable, and risk the fossil fuel industry exploiting it so that it can continue to pollute? It's a difficult dilemma. My personal view is that I can't change the laws of physics, so I can't ignore drawdown. But I might be able to influence the fossil fuel industry, so the risk of them taking advantage of drawdown is the lesser of two evils.

I believe that within a decade humanity will be desperately scrambling for enduring solutions to the climate emergency. Our situation will be a bit like the initial containment phase of the COVID-19 pandemic, in that we will not see the benefits of the actions we have taken immediately, nor will we have the long-term solution of a vaccine (or drawdown) enabling a return to 'normal'.

The ultimate solution to the COVID-19 problem is a vaccine. But the development of vaccines is expensive and takes time. The search for a vaccine for COVID-19 is likely to cost many billions. And there's no certainty that scientists will succeed. The comparison with climate change is stark. If we are to maximise our chances of staying safe from climate impacts, we will need to invest billions in the development of a 'climate vaccine' consisting of technologies that can take the greenhouse gas CO_2 out of the air.

Drawdown is a goal for the medium term. But if it is to reach its full potential we need to make a very large investment in it right now. And we must embark on this project regardless of whether it is possible to develop a sufficient solution in time to save us.

So how do the drawdown pathways work? Photosynthesis is one example. Before a tree (or any plant) grows, the carbon that will make it is in the atmosphere, as the gas carbon dioxide. The tree takes in CO_2 and uses energy from the sun to power photosynthesis, which breaks apart the CO_2 so the carbon in it can be used to build plant tissue, and the oxygen is expelled.

You can think of a tree as congealed CO_2, which is safely stored for as long as the tree lives. But after the tree dies and rots, or is burned, all of the carbon in it becomes CO_2, which returns to the atmosphere. But some trees are buried and ultimately turned into coal, which takes the carbon in them out of the climate system for millions of years. While using very different processes, the best drawdown options have the same effect as turning trees into coal.

It's hard finding that much land for trees when agriculture, cities, roads and farms take up so much of Earth's surface. In fact, just protecting our existing forests is a challenge. They are threatened by drought, insect attack, land-clearing and fire. The

Australian bushfires of 2019–20 offer an example of how severe the destruction of our forests is: 21% of Australia's broad-leafed temperate forests burned in those megafires, and it will take many years for all the carbon released to be recaptured by new trees.

There are, in my view, only four realistic possibilities to draw down carbon at the scale required. And all are at a very early stage of development.

SEAWEED

The oceans cover 71% of the Earth's surface, and there is little competition for space in the open ocean. Seaweeds are also among the fastest-growing plants, which means they draw carbon out of the atmosphere at a faster rate than other plants. There's a whole atmosphere worth of gas dissolved in the oceans, and the gases transfer readily between this dissolved atmosphere and the atmosphere we breathe. Seaweed grows by drawing CO_2 out of the dissolved atmosphere and, as it does so, it makes the oceans less acid, which has great benefits for marine life. If the seaweed is then sunk into the deep ocean, the carbon in it can be kept from returning to the atmosphere.

If we covered 9% of the world's oceans in seaweed, we could draw down the equivalent of all the greenhouse gases we currently emit. That would take the development of an area of seaweed farms, in the open ocean, covering an area about 4.5 times the size of Australia. Today, there is not a single seaweed farm in deep water (where the carbon could sink to safe depths) anywhere on Earth. If we work very hard now, researching the basic science, and investing in early stage development, the best I think we could achieve by 2050 is seaweed farms equal in area to South Australia, which would draw down about one gigatonne of CO_2 per year. But even that would be a large contribution to avoiding tipping points.

The oceans also offer a place to store the carbon we draw from the atmosphere. If seaweed drifts down to depths below 1000 metres, as it often does naturally, the carbon in it remains locked away and won't return to the atmosphere for hundreds of years or even millennia. That's because the ocean water circulation is slow, and gas is not easily exchanged between the ocean's deeper layers and its surface layers, meaning that the CO_2 and carbon sunk to the depths can't easily rise.[3]

So exactly what needs to be done if seaweed is to provide a scalable drawdown tool in coming decades? For a start, we need more research. We don't know, for example, whether by dropping seaweed into the ocean depths by the gigatonne, we will harm life there. But we do know that hundreds of thousands of tonnes of seaweed reach the deep sea naturally. And that the deep sea is so vast that we could sink half of the atmosphere's CO_2 into it, and yet only increase the concentration of CO_2 in the deep ocean by 2%. As we experiment to determine the answer to such questions, we must be driven by the inevitability of what will transpire if we do nothing.

If we are to fast track the use of seaweed for drawdown, we'll need a map of the oceans showing where we might begin growing seaweed to sequester carbon. Such a map would need to take into account the location of submarine canyons, which can provide the required depths, and proximity to port facilities and nutrients, and the necessary technology. It's telling of the state of development of drawdown that even such basics are lacking.

The engineering challenges of growing seaweed in the open ocean are enormous and barely explored, as are the economic and aquaculture challenges. The Australian Government could take a global lead in the seaweed-for-drawdown space by hosting a summit on the potential of seaweed to address climate change. Based on the summit's findings, a flagship organisation could be established in the CSIRO or a university, fleshing out the steps to making seaweed sequestration a reality.

There is also a role for philanthropy in this space. If government won't convene a summit, philanthropists could. And philanthropy could assist with the work required to fill in the many gaps we face in understanding seaweed as a drawdown tool. Entrepreneurs can also help. One example of a young entrepreneur doing this is former fashion designer Sam Elsom. He abruptly changed course and has started growing seaweed to protect the climate.

Seaweed, which is already being grown for many purposes, is a booming industry, and more uses are being discovered for it by the week. If entrepreneurs help create a robust seaweed industry in Australia, even if the seaweed is not sequestered, we may make major breakthroughs along the way.

SILICATE ROCKS

Silicate rocks sequester CO_2 as they weather. These rocks have been a major component in maintaining the balance of atmospheric CO_2 over the ages. Australia is exceptionally rich in silicate rocks, and has unsurpassed experience in mining, making it a potential leader in the field of silicate rock drawdown.

A lot of scientific research has gone into understanding how much silicate rocks could contribute to drawdown. James Hansen, one of the world's leading climate scientists, thinks that silicate rocks could draw all of the excess CO_2 out of the atmosphere by 2100, if we could mine enough of the rocks.[4] Options for use of the mined, crushed rocks include using them to rejuvenate beaches being lost to rising seas, as a soil amendment in agriculture, and even in dropping them into the deep ocean. In all of these places they would decompose, capturing CO_2.

The most obvious problem with using silicate rocks to drawdown CO_2 is that, in the current, polluting economy, a lot of fossil fuel is burned in the processes of mining, crushing and transporting them. This means that we'd lose much, if not all, of the gains we'd get from using silicate rocks for drawdown. We could be mining and transporting silicate rocks using emissions-free technology in a decade, so we have ten years to learn how we can best utilise silicate rocks for drawdown, and in ways that do least harm to the environment. Once we've done the required research and planning, we'd have the clean energy required to start using silicate rocks at scale.

The other major problem with the use of silicate rocks for drawdown is the question of who would pay for the service. The Federal Government, under Tony Abbott, destroyed Australia's only chance at instating a carbon price, and without a carbon price, there is no obvious purchaser of the service provided by drawdown using silicate rocks, as there is no product except less CO_2 in the atmosphere. It seems inevitable that a global carbon price will be imposed eventually. If and when it is, the motive force of drawdown using silicate rocks and seaweed will be unleashed.

Australia is a world leader in mining. It is arguably the best-positioned of any nation to lead research into how silicate rocks might be used to help avert the climate emergency.

Whether funded by industry, philanthropy or government, Australia would be well served by establishing a Centre for Drawdown Research, charged with looking into various drawdown options, including using silicate rocks. It could conduct the basic science required, look into economic and environmental issues, and start trialling approaches. Such a centre could be seen as some compensation to the world for the disgraceful role Australia has played in stymying effective global climate action in recent years.

CARBON-NEGATIVE BUILDING MATERIALS

Cities are responsible for 70% of global emissions, much of which comes from construction.[5] Yet we already have the means to make construction emissions free. One of the largest single sources of emissions is the production of concrete, which is responsible for around 8% of the world's annual CO_2 emissions (about twice that of air travel).[6]

One study claims that the production of 'cement can remove more CO_2 than it adds to the atmosphere'.[7] This is because, 'after manufacture cement undergoes a 'carbonation' process wherein concrete exposed to CO_2 and humidity slowly bonds with the CO_2, storing the carbon in mineral form. Because concrete is porous, CO_2 can slowly diffuse into concrete, carbonating the cement to a depth of 60 millimetres or more over a number of years.' If the concrete is made without producing CO_2 emissions (for example, by using fly ash from the dumps of old coal-fired power plants), the overall effect (ignoring the coal burned to create the ash) is drawdown of CO_2.

An abundance of other carbon negative opportunities exist in the construction sector, from using timber in long-lived structures, to the planting of urban forests. But the opportunities to reduce emissions in cities is even greater. This is a subject beyond the scope of this book, but it's clear that one of the most critical arenas in which our climate future will be decided is the future form of our cities.

DIRECT AIR CAPTURE OF CO_2

There are several methods for drawing CO_2 directly out of the atmosphere that are already in commercial use. A common method of direct air capture (DAC) involves the use of an alkaline liquid, but resins can also absorb the CO_2. The CO_2 is released from the resin when it is heated. It can then be transported and sequestered.

The two leading commercial companies in the field are Canadian-based Carbon Engineering and the Swiss Climeworks. Carbon Engineering uses the CO_2 to manufacture (among other products) e-fuel. It has claimed that it could geologically sequester the CO_2 it captures for about US$100 per tonne. Climeworks uses the CO_2 it captures to feed into greenhouses to enhance plant growth, and to carbonate soft drinks. In 2017 Carbon Engineering entered a partnership with Reykjavik Energy to establish a project called CarbFix, which uses low-grade waste heat from the plant to run the process, and to inject the captured CO_2 700 metres into the ground, where it reacts with the rocks and becomes a solid carbonate.

As interesting as direct air capture is as a process, the fact that it requires large machines, along with the costs involved in sequestration, means that the opportunities for large-scale operations are limited. Nonetheless, if the opportunity is grasped, it would not be surprising if DAC was drawing one gigatonne of CO_2 from the air per year by 2050.

With so many opportunities to contribute, it's pathetic that, so far, Australia's sole contribution to drawdown has been the Emissions Reduction Fund. But because it's counted in our national carbon accounts, it currently is used as justification for continuing to burn fossil fuels. We can do so much better than this if we harness our innovative thinking and the vast natural potential of our lands and seas.

We live in a world where climate change threatens our economies, our land and our lives. The seas are rising, largely unnoticed; birds, frogs and other animals are silently disappearing. The summers are getting warmer and drier. We have sleepwalked deep into the world that exists just seconds before the climate clock strikes a catastrophic midnight. But finally Australians are waking to the seriousness of the threat we face.

And at the last moment, between megafires and COVID-19,

governments are at last getting serious about the business of governance. We know that we can change as individuals, and also that governments can act decisively. As we turn to combatting the climate emergency, we know that we can build on the huge amount that has already been achieved. The groundwork has already been laid, to drive fossil fuels from our energy systems. And the Federal Government is returning to consideration of adaptation measures, such as beginning to plan to save the Great Barrier Reef. But we must see all of this as just a beginning in a battle for survival that will continue for decades. So very much needs to be done. We cannot afford to waste a single day as we build the capacity to face an uncertain future.

The next few years are our last chance to act.

Actions Summary

The following list of new institutions, policies and actions is my best effort at envisaging what is required for Australia to survive the climate emergency.

- A National Target and Plan for 95% or more of electricity to be supplied by renewables by 2030.
- State plans to electrify all transport, beginning with the swift retirement of non-electric buses and including a plan for 50% of all new car sales to be EVs by 2030.
- Implement planned changes to how we work and live so as to minimise unnecessary travel.
- A plan for clean hydrogen to replace bunker fuel in shipping.
- A plan for the adoption of e-fuels for aviation, with an aim to have all domestic flights running on e-fuel by 2030.
- A National Commission for Climate Adaptation, with a Coastal Defence Fund and a Commission for Primary Production operating under its umbrella.
- A National Initiative on Drawdown Innovation to provide leadership in early stage research and fund some on-ground projects.
- The Federal Government to help convene a Global Working Group on Geoengineering.

Acknowledgments

Great thanks to Andrew Stock for a thorough review with special emphasis on technical and engineering aspects. Thanks also to Professor Will Steffen, whose thorough read through with special emphasis on climate change and Earth systems aspects assisted me greatly. Sam MacLean also assisted me with the chapter dealing with EVs.

Michael Heyward and Jane Pearson read the manuscript and greatly improved it with their editorial suggestions. As always, great thanks to Kate Holden for encouraging me in my work, and to both her and Coleby Holden for helping create the time for me to do it.

Notes

Introduction

1 Morgan, E. and Long, S., 'Coronavirus Economic Recovery Committee Looks Set to Push Australia towards Gas-fired Future', ABC News, 13 May 2020, https://www.abc.net.au/news/2020-05-13/coronavirus-recovery-to-push-australia-towards-gas-future/12239978

2 Lenton, T. *et al*, 'Climate Tipping Points—Too Risky to Bet Against', *Nature*, 575, 592–95, 9 April 2020, https://www.nature.com/articles/d41586-019-03595-0

3 Brendryen, J., Haflidason, H., Yokoyama, Y. *et al*, 'Eurasian Ice Sheet Collapse Was a Major Source of Meltwater Pulse 1A 14,600 Years Ago', *Nature Geoscience*, 20 April 2020, https://www.nature.com/articles/s41561-020-0567-4

4 Lenton, T. *et al*, 'Climate Tipping Points—Too Risky to Bet Against', *Nature* 575, 592–95, 9 April 2020, https://www.nature.com/articles/d41586-019-03595-0

5 *Ibid*.

6 Known as the Council of Parties meeting no. 25, abbreviated to COP 25.

7 Nguyen, K. *et al*, 'The Truth about Australia's Fires—Arsonists Aren't Responsible for Many this Season', ABC News, 18 January 2020, https://www.abc.net.au/news/2020-01-11/australias-fires-reveal-arson-not-a-major-cause/11855022

8 Morton, A., 'Hazard Reduction Burning Had Little to No Effect on Slowing Extreme Bushfires', *Guardian*, 6 February 2020, https://www.theguardian.com/environment/2020/feb/06/hazard-reduction-burning-had-little-to-no-effect-in-slowing-this-summers-bushfires

9 Grattan, M., 'View from the Hill: Scott Morrison Announces Mandatory Self-isolation for All Overseas Arrivals and Gives Up Shaking Hands', *The Conversation*, 15 March 2020, http://theconversation.com/view-from-the-hill-scott-morrison-announces-mandatory-self-isolation-for-all-overseas-arrivals-and-gives-up-shaking-hands-133715

10 Blau, A., 'What Australians Really Think about Climate Action', ABC News, 5 February 2020, https://www.abc.net.au/news/2020-02-05/australia-attitudes-climate-change-action-morrison-government/11878510?nw=0

11 Daugy, M., 'Key Electricity Trends 2019', IEA, 14 April 2020, https://www.iea.org/articles/key-electricity-trends-2019

12 Carbon Tracker Press Release, '42% of Global Coal Power Plants Run at a Loss, Finds World-first Study', Carbon Tracker, 30 November 2018,

https://carbontracker.org/42-of-global-coal-power-plants-run-at-a-loss-finds-world-first-study/

13 *Ibid.*

14 Subsidies include the federal fuel tax credits scheme, accelerated depreciation, and concessional rates of excise on aviation fuel.

15 Stock, A. *et al*, 'End of the Line: Coal in Australia', Climate Council of Australia Report, 2018, https://www.climatecouncil.org.au/wp-content/uploads/2018/07/CC_MVSA0148-Report-End-of-the-Line-Coal_V3-FA_Low-Res_Single-Pages.pdf

16 Hausfather, Z., 'Analysis: Global Fossil Fuel Emissions up 0.6% in 2019 Due to China', *Carbon Brief*, 4 December 2019, https://www.carbonbrief.org/analysis-global-fossil-fuel-emissions-up-zero-point-six-per-cent-in-2019-due-to-china#:~:text=Emissions%20from%20fossil%20fuel%20and,Global%20Carbon%20Project%20(GCP). Rice, D., 'Global Carbon Dioxide Emissions Stayed Flat in 2019, Despite Growing Economy', USA Today, 12 February 2020, https://www.usatoday.com/story/news/world/2020/02/12/climate-change-global-carbon-dioxide-emissions-stayed-flat-2019/4737218002/

17 Hill, J., 'Biggest Drop in CO_2 Emissions in 30 Years in 2019—but It's Not Nearly Enough', *RenewEconomy*, 11 March 2020, https://reneweconomy.com.au/biggest-drop-in-co2-emissions-in-30-years-in-2019-but-its-not-nearly-enough-24403/

18 McIntosh, E., 'Teck Resources Pulls Application for Frontier Oilsands Mine, Citing Need for Climate Action', *National Observer*, 23 February, 2020, https://www.nationalobserver.com/2020/02/23/news/teck-resources-pulls-application-frontier-oilsands-mine-citing-need-climate-action

19 Healing, D., 'Dozens of Alberta Oilsands Projects Won't Be Built in the Near Future, Analysts Say', *Global News*, 1 March, 2020, https://globalnews.ca/news/6615579/alberta-oilsands-teck-frontier-analysis/

20 ABC News, 'Equinor Abandons Plans to Drill for Oil in the Great Australian Bight for "Commercial" Reasons', 25 February 2020, https://www.abc.net.au/news/2020-02-25/equinor-abandons-plan-to-drill-for-oil-in-great-australian-bight/11997910

21 Blackmon, D., 'BP's Big Writedown: A Harbinger for a Declining Industry, or of a Struggling Company?', *Forbes*, 16 June 2020, https://www.forbes.com/sites/davidblackmon/2020/06/16/bps-big-writedown-a-harbinger-for-a-declining-industry-or-of-a-struggling-company/#6abb1c372d46

22 Emissions intensity is a measure of emissions relative to production. 'Rio Tinto to Invest $1 Billion to Help Meet New Climate Change Targets', Rio Tinto, 26 February 2020, https://www.riotinto.com/en/news/releases/2020/

Rio-Tinto-to-invest-1-billion-to-help-meet-new-climate-change-targets

23 Vorrath, S., 'Mining Giants BHP, Anglo and Fortescue Join Forces for "Green Hydrogen"', RenewEconomy 18 March 2020, https://reneweconomy.com.au/mining-giants-bhp-anglo-and-fortescue-join-forces-for-green-hydrogen-48061/

Chapter 1

1 Karp, P., 'Morrison Says No Evidence Links Australia's Carbon Emissions to Bushfires', *Guardian*, 27 November, 2019, https://www.theguardian.com/australia-news/2019/nov/21/scott-morrison-says-no-evidence-links-australias-carbon-emissions-to-bushfires

2 'Australia's Low Pollution Future: The Economics of Climate Change Mitigation', Australian Government, Treasury, 30 October 2008, https://web.archive.org/web/20100718083633/http:/www.treasury.gov.au/lowpollutionfuture/

3 The webpage detailing the Treasury modelling has been deleted.

4 This is an estimate. In 2017 it was half of all Liberal members in the Federal Parliament (see Patrick, A., 'More than Half of Federal Liberal MPs "Don't Trust" Climate Science: Think Tank', *Australian Financial Review*, 17 July 2017, https://www.afr.com/policy/energy-and-climate/more-than-half-of-federal-liberal-mps-dont-trust-climate-science-think-tank-20170714-gxb7r2). Following the 2019 federal election and the 2019–20 megafires, that number has dropped.

5 Hannam, P., '"Like Terrorists": Malcolm Turnbull Assails Liberal Climate Deniers', *Sydney Morning Herald*, 6 February 2020, https://www.smh.com.au/politics/federal/like-terrorists-malcolm-turnbull-assails-liberal-climate-deniers-20200206-p53y6u.html

6 Franta, B., 'Shell and Exxon's Secret 1980s Climate Change Warnings', *Guardian*, 19 September 2018, https://www.theguardian.com/environment/climate-consensus-97-per-cent/2018/sep/19/shell-and-exxons-secret-1980s-climate-change-warnings

7 Figueres, C. and Rivett-Carnac, T., *The Future We Choose*, Bonnier, London, 2020.

8 Mazengarb, M., 'Global Fossil Fuel Subsidies Reach $5.2 Trillion, and $29 Billion in Australia', *Renew Economy*, 13 May 2019, https://reneweconomy.com.au/global-fossil-fuel-subsidies-reach-5-2-trillion-and-29-billion-in-australia-91592/. Climate Council, 'Air Quality: What You Need to Know', 14 January 2020, Climatecouncil.org.au/resources/air-quality/

9 AIHW [Australian Institute of Health and Welfare], 'Australian Burden of Disease Study: Impact and Causes of Illness and Death in Australia

2011', Canberra, 10, May 2016, https://www.aihw.gov.au/reports/burden-of-disease/abds-impact-and-causes-of-illness-death-2011/contents/highlights

Chapter 2

1 Readfearn, G., 'UK Climate Scientist Corrects Australian MP Craig Kelly's "Blatant Misrepresentation"', *Guardian*, 22 January 2020, https://www.theguardian.com/environment/2020/jan/22/uk-climate-scientist-corrects-australian-mp-craig-kellys-blatant-misrepresentation
2 Clean Energy Council, 'Renewable Energy Target', https://www.cleanenergycouncil.org.au/advocacy-initiatives/renewable-energy-target
3 Australian Government, Department of Industry, Science, Energy and Resources, 'Estimating Greenhouse Gas Emissions from Bushfires in Australia's Temperate Forests: Focus on 2019–20', April 2020, https://www.industry.gov.au/data-and-publications/estimating-greenhouse-gas-emissions-from-bushfires-in-australias-temperate-forests-focus-on-2019-20
4 Warner, J. and Lyons, S., 'The Size of Australia's Bushfire Crisis Captured in Five Big Numbers', ABC News, 5 March 2020, https://www.abc.net.au/news/science/2020-03-05/bushfire-crisis-five-big-numbers/12007716
5 Morton, A., 'A 60% Rise in Industrial Emissions Points to Failure of Coalition's "Safeguard Mechanism"', *Guardian*, 12 February 2020, https://www.theguardian.com/australia-news/2020/feb/12/a-60-rise-in-industrial-emissions-points-to-failure-of-coalitions-safeguard-mechanism
6 AEMC, 'Five Minute Settlement', https://www.aemc.gov.au/rule-changes/five-minute-settlement#:~:text=On%2028%20November%202017%20the,five%20minutes%2C%20starting%20in%202021.&text=Over%20time%2C%20this

Chapter 3

1 Grose, M. and Arblaster, J., 'Just How Hot Will It Get this Century? Latest Climate Models Suggest It Could Be Worse than We Thought', *The Conversation*, 18 May 2020, https://theconversation.com/just-how-hot-will-it-get-this-century-latest-climate-models-suggest-it-could-be-worse-than-we-thought-137281
2 *Ibid.*
3 Steffen, W. and Hughes, L., 'The Critical Decade 2013: Climate Change Science, Risks and Responses', Climate Council, June 2013, http://www.climatecouncil.org.au/uploads/b7e53b20a7d6573e1ab269d36bb9b07c.pdf
4 Australian Government, Department of Industry, Science, Energy and Resources, 'National Greenhouse Gas Inventory: December

2019', https://www.industry.gov.au/data-and-publications/national-greenhouse-gas-inventory-december-2019

5 Imster, E. and Byrd, D., 'Atmospheric Concentration of CO_2 Hits Record High in May 2019', *EarthSky*, 17 June 2019, https://earthsky.org/earth/atmospheric-co2-record-high-may-2019

6 Measured as CO_2 equivalent.

7 Berwin B., 'Fossil Fuel Emissions Push Greenhouse Gas Indicators to Record High in May', *Inside Climate News*, 5 June 2020, https://insideclimatenews.org/news/04062020/fossil-fuel-emissions-mauna-loa-keeling-curve-coronavirus-hawaii

8 Mann, M. E., 'It's Not Rocket Science: Climate Change Was behind this Summer's Extreme Weather', *Washington Post*, 3 November 2018.

9 Schurer, A. P. *et al*, 'Interpretations of the Paris Climate Target', *Nature Geoscience*, 19 March 2018, http://www.meteo.psu.edu/holocene/public_html/Mann/articles/articles/SchurerEtAlNatureGeosci18.pdf

10 Mann, M. E., Presentation to Australian Museum, 24 March 2020.

Chapter 4

1 *Picturesque Atlas of Australia*, 1856.

2 Technically it's broad-leaf temperate forests.

3 Cox, L., '"Unprecedented" Globally: More than 20% of Australia's Forests Burned in Bushfires', *Guardian*, 25 February 2020, https://www.theguardian.com/australia-news/2020/feb/25/unprecedented-globally-more-than-20-of-australias-forests-burnt-in-bushfires. Nature Climate Change, 'In The Line of Fire, 24 February 2020, https://www.nature.com/articles/s41558-020-0720-5

4 Hennessy, K. *et al*, 'Climate Change Impacts on Fire-weather in South-east Australia', Consultancy report for the New South Wales Greenhouse Office, Victorian Department of Sustainability and Environment, ACT Government, Tasmanian Department of Primary Industries, Water and Environment and the Australian Greenhouse Office. CSIRO Marine and Atmospheric Research and Australian Government Bureau of Meteorology, 2005, http://www.cmar.csiro.au/e-print/open/hennessykj_2005b.pdf.

5 Bureau of Meteorology, 'Annual Climate Statement 2019', http://www.bom.gov.au/climate/current/annual/aus/#:~:text=2019%20was%20Australia's%20warmest%20year%20on%20record%2C%20with%20the%20annual,1.33%20%C2%B0C%20in%202013

6 Readfearn, G., '2019 Was Australia's Hottest Year on Record—1.5C above Average Temperature', *Guardian*, 2 January 2020, https://www.theguardian.com/australia-news/2020/

jan/02/2019-australia-hottest-year-record-temperature-15c-above-average-temperature

7 Lewis, S. *et al*, 'Quantitative Estimates of Anthropogenic Contributions to Extreme National and State Monthly, Seasonal and Annual Average Temperatures for Australia', *Australian Meteorological and Oceanographic Journal*, April 2014, http://www.bom.gov.au/jshess/docs/2014/lewis2.pdf

8 Australian Government, Bureau of Meteorology, Annual Climate Statement 2018, 10 January 2019, http://www.bom.gov.au/climate/current/annual/aus/2018/

9 Werner, J. and Lyons, S., 'The Size of Australia's Bushfire Crisis Captured in 5 Big Numbers.', ABC News, 5 March 2020, https://www.abc.net.au/news/science/2020-03-05/bushfire-crisis-five-big-numbers/12007716

10 Professor Fay Johnson's testimony to the bushfire royal commission, 26 May 2020, reported in the *Guardian*, https://www.theguardian.com/australia-news/2020/may/26/australias-summer-bushfire-smoke-killed-445-and-put-thousands-in-hospital-inquiry-hears#:~:text=Smoke%20from%20the%20Australian%20bushfires,six%20states%20across%20six%20months.)

11 Lucas, C. *et al*, 'A Crisis of Underinsurance Threatens to Scar Rural Australia', *The Conversation*, 7 January 2020, https://theconversation.com/a-crisis-of-underinsurance-threatens-to-scar-rural-australia-permanently-129343

12 McCormack, L., 'Opinion: The Human Cost—Psychological Impact of the Bushfires', University of Newcastle, 30 March 2020, https://www.newcastle.edu.au/newsroom/featured/opinion-the-human-cost-psychological-impact-of-the-bushfires

13 'Tura Beach Mother Jen Spears Fears Bushfire Smoke Impact on her Unborn Baby's Health', ABC News, 6 August 2020, https://www.abc.net.au/news/2020-08-06/fears-of-impact-of-bushfire-smoke-on-unborn-babies/12528180

14 Lowrey, T., 'Tathra Bushfires Anniversary Brings Another Financial Blow for Many Already Suffering', ABC News, 18 March 2019, https://www.abc.net.au/news/2019-03-18/tathra-fires-anniversary-means-end-of-rent-insurance/10909800. Omar Khalifa, pers. comm.

15 Julian Armstrong, Protection Supervisor, South Coast Forestry Corporation of NSW, email to Omar Khalifa, University of Wollongong.

16 Ferguson, K. and Henderson, K., 'Bushfire Grants and Loans Face Shakeup Over Lengthy Application Process', ABC News, 2 March 2020, https://www.abc.net.au/news/2020-03-02/government-concedes-bushfire-grants-taking-too-long/12018560.

Chapter 5

1 Reed, P. and Dennis, R., 'With Costs Approaching $100 Billion, the Fires Are Australia's Costliest Natural Disaster', *The Conversation*, 17 January 2020, https://theconversation.com/with-costs-approaching-100-billion-the-fires-are-australias-costliest-natural-disaster-129433

2 Well, D., 'Climate Change Named Biggest Global Threat in New WEF Risks Report', *EcoWatch*, 15 January 2020, https://www.ecowatch.com/climate-change-global-risk-wef-2644819254.html

3 'Resilience to Climate Change?', *Economist*, https://www.eiu.com/public/topical_report.aspx?campaignid=climatechange2019&zid=climatechange2019&utm_source=blog&utm_medium=one_site&utm_name=climatechange2019&utm_term=announcement&utm_content=bottom_link

4 Partington, R. J., 'G20 Sounds Alarm over Climate Emergency Despite US Objections', *Guardian*, 24 February 2020, https://www.theguardian.com/world/2020/feb/23/g20-sounds-alarm-over-climate-emergency

5 Sullivan, K., 'Climate Change Slashes More than $1 Billion from Farm Production Value Over Past 20 Years: ABARES', ABC News, 18 December 2019, https://www.abc.net.au/news/rural/2019-12-18/abares-climate-change-slashes-1-billion-farm-incomes-20-years/11809500

6 Bell, S. J., 'Farmers Impacted by Bushfires Count "Heartbreaking" Cost as Livestock Losses Climb', ABC Ballarat News, 7 January 2020, https://www.abc.net.au/news/2020-01-07/farmers-recount-heartbreaking-toll-of-bushfire-livestock-losses/11844696

7 Kurmelovs, R., 'No Vintage: Australian Vinyards Dump Grape Harvest as Bushfire Smoke Takes Its Toll', *Guardian*, 8 February 2020, https://www.theguardian.com/australia-news/2020/feb/07/no-vintage-australian-vineyards-dump-grape-harvest-as-bushfire-smoke-takes-its-toll

8 David Karoly, pers. comm. Based on Lewis, S. *et al*, 'Quantitative Estimates of Anthropogenic Contributions to Extreme National and State Monthly, Seasonal and Annual Average Temperatures for Australia', *Australian Meteorological and Oceanographic Journal* 64, 2014, pp. 215–30.

9 Readfearn, G., 'Great Barrier Reef Could Face "Most Extensive Coral Bleaching Ever" scientists say', *Guardian*, 22 February 2020, https://www.theguardian.com/environment/2020/feb/22/great-barrier-reef-could-face-most-extensive-coral-bleaching-ever-scientists-say

10 Hughes, T. P. *et al*, 'Spatial and Temporal Patterns of Mass Bleaching of Corals in the Anthropocene', *Science*, 359, (6371), January 2018, pp. 80–83, https://science.sciencemag.org/content/359/6371/80

11 Gibbens, S., 'Scientists Are Trying to Save Coral Reefs. Here's What's Working', *National Geographic Science*, 4 June

2020, https://www.nationalgeographic.com/science/2020/06/scientists-work-to-save-coral-reefs-climate-change-marine-parks/

12 'Great Barrier Reef: Bleaching Could Cost Queensland $1 Billion Annually', Climate Council, 20 April 2017, https://www.climatecouncil.org.au/resources/media-release-great-barrier-grief-bleaching-could-cost-queensland-1-billion-annually/

13 Church, J. and Zhang, X., '15 Years from Now, Our Impact on Regional Sea-level Will Be Clear', *The Conversation*, 13 October 2014, https://theconversation.com/15-years-from-now-our-impact-on-regional-sea-level-will-be-clear-31821

14 http://coastalrisk.com.au/viewer

15 Weeman, K., 'New Study Finds that Sea Level Rise Is Accelerating', NASA Global Climate Change, 13 February 2018, https://climate.nasa.gov/news/2680/new-study-finds-sea-level-rise-accelerating/

16 Milman, O., 'Greenland's Melting Ice Raised Global Sea Level by 2.2mm in Two Months', *Guardian*, 19 March 2020, https://www.theguardian.com/science/2020/mar/19/greenland-ice-melt-sea-level-rise-climate-crisis

17 Witze, A., 'Why Extreme Rains Are Gaining Strength as the Climate Warms', *Nature*, 22 November 2018. https://www.nature.com/articles/d41586-018-07447-1

18 'How Much Will Sea Levels Rise in the 21st Century?', *Skeptical Science*, 12 July 2015, https://skepticalscience.com/sea-level-rise-predictions.htm

19 Fairhall, L., 'Walter Waia Is a Culture Man', *Dumbo Feather*, Fourth Quarter 2019.

20 *Ibid*.

21 Preiss, B., 'Mount Martha Beach and Its Beach Boxes Cannot Be Saved, Government Concedes', *Age*, 16 August 2019, https://www.theage.com.au/national/victoria/mount-martha-beach-and-its-beach-boxes-cannot-be-saved-government-concedes-20190816-p52hyd.html

22 Burgess, G., 'Tasmanian Councils Grapple with Effects and Price Tag of Climate Change', ABC Radio Hobart, 5 June 2019, https://www.abc.net.au/news/2019-06-05/this-is-what-climate-change-looks-like-in-tasmania/11176182

23 Ceranic, I. *et al*, 'A Coast Being Slowly Eaten by the Ocean', ABC News, 5 August 2019, https://www.abc.net.au/news/2019-07-31/erosion-washing-away-beaches-up-and-down-wa-coast/11359006

24 Page, D., 'City of Newcastle Closes Stockton Beach as Relentless Conditions Cause Further Devastating Erosion', *Newcastle Herald*, 19 September 2019, https://www.newcastleherald.com.au/story/6393946/watch-the-video-stockton-on-the-edge-as-erosion-threatens-caravan-park-and-cafe/

25 Page, D., 'Resident's Desperate Plea for Help as Stockton's

Identity Erodes with the Beach', *Newcastle Herald,* 20 September 2019, https://www.newcastleherald.com.au/story/6395059/stocktons-sos-sand-loss-so-great-the-beach-is-broken/

26 Vousdoukas, M. *et al*, 'Sandy Coastlines under Threat of Erosion', *Nature Climate Change*, 2 March 2020, https://www.nature.com/articles/s41558-020-0697-0

27 Steffen, W. *et al*, 'Counting the Costs: Climate Change and Coastal Flooding', Climate Council, 2014, https://www.climatecouncil.org.au/uploads/56812f1261b168e02032126342619dad.pdf

28 Steffen, W. *et al*, 'Compound Costs: How Climate Change Is Damaging Australia's Economy', Climate Council, 2019, https://www.climatecouncil.org.au/wp-content/uploads/2019/05/costs-of-climate-change-report-v2.pdf

Chapter 6

1 Phillips, T., 'Jair Bolsonaro Claims Brazilians "Never Catch Anything" as COVID-19 Cases Rise', *Guardian*, 27 March 2020, https://www.theguardian.com/global-development/2020/mar/27/jair-bolsonaro-claims-brazilians-never-catch-anything-as-covid-19-cases-rise

2 'Brazil: Bolsonaro Sabotages Anti-COVID-19 Efforts', *Human Rights Watch*, 10 April 2020, https://www.hrw.org/news/2020/04/10/brazil-bolsonaro-sabotages-anti-covid-19-efforts

3 'Covid-19 in Brazil: "So What?"', *Lancet*, 395, 9 May 2020, https://www.thelancet.com/pdfs/journals/lancet/PIIS0140-6736(20)31095-3.pdf

4 Phillips, D., 'Brazil Stops Releasing COVID-19 Death Toll and Wipes Data from Official Site', *Guardian*, 8 June 2020, https://www.theguardian.com/world/2020/jun/07/brazil-stops-releasing-covid-19-death-toll-and-wipes-data-from-official-site. Johns Hopkins University & Medicine, Coronavirus Resource Center, 16 July 2020, https://coronavirus.jhu.edu/map.html

5 Maisch, M., 'Climate Policy Done Right: Tasmania Sets a 200% RE Target by 2040', *PV Magazine*, 5 March 2020, https://www.pv-magazine-australia.com/2020/03/05/climate-policy-done-right-tasmania-sets-200-re-target-by-2040/. Gramenz, E., 'Tasmania's Eye on Full Renewable Electricity by 2022, but Work Still to Be Done on Securing Supply', ABC News, 16 August 2017, https://www.abc.net.au/news/2017-08-16/recommendations-from-tas-energy-security-taskforce/8814220

6 Vorath, S., 'South Australia on Track to 100 pct Renewables as Regulator Comes to Party', *RenewEconomy*, 24 January 2020, https://reneweconomy.com.au/south-australia-on-track-to-100-pct-renewables-as-regulator-comes-to-party-96366/

7 Schmidt, B., 'NSW Net Zero Plan Lays out Path for More Electric Vehicles', *The Driven*, 16 March 2020, https://thedriven.io/2020/03/16/

nsw-net-zero-plan-lays-out-path-for-more-electric-vehicles/

8 Cities Power Partnership, https://citiespowerpartnership.org.au/

9 Australian Government, Australian Renewable Energy, 'Solar Energy', https://arena.gov.au/renewable-energy/solar/

10 Parkinson, G., 'Australia Rooftop Solar Installs Total 2.13GW in 2019 after Huge December Rush', *RenewEconomy*, 12 January 2020, https://reneweconomy.com.au/australia-rooftop-solar-installs-total-2-13gw-in-2019-after-huge-december-rush-34613/

Chapter 7

1 *Australian Mining Review*, March 2020, http://www.australianminingreview.com.au/pdf/AMR-MARCH-20.pdf

2 Stern, N. H., *The Economics of Climate Change: The Stern Review*, Cambridge University Press, Cambridge, UK, 2007.

3 Parkinson, G., 'Solar, Wind and Battery Storage Now Cheapest Energy Option Just about Everywhere', *RenewEconomy*, 28 April 2020, https://reneweconomy.com.au/solar-wind-and-battery-storage-now-cheapest-energy-options-just-about-everywhere-95748/

4 Mazengarb, M., 'Electricity Prices Set to Plummet as New Wind and Solar Kick in', *RenewEconomy*, 9 December 2019, https://reneweconomy.com.au/electricity-prices-set-to-plummet-as-strong-wind-and-solar-investment-kicks-in-77816/

5 'Breaking It Down: Victorian Coal Power Plants the Least Reliable in Aus', Australia Institute, 12 February 2020, https://www.tai.org.au/content/breaking-it-down-victorian-coal-power-plants-least-reliable-aus

6 The average over the thirteen years was 6.47 TWH, the lowest usage was 4.5 in both 2015 and 2016. Data from Open.Nem. Provided by Andrew Stock.

7 *Ibid.*

8 Harmsen, N., 'South Australia's Giant Tesla Battery Output and Storage Set to Increase by 50%', ABC News 19 November 2019, https://www.abc.net.au/news/2019-11-19/sa-big-battery-set-to-get-even-bigger/11716784

9 ACT Government, Environment, Planning and Sustainable Development Directorate, https://www.environment.act.gov.au/cc/act-climate-change-strategy/emission-reduction-targets

10 Marzengarb, M., 'Evoenergy Turns to Renewable Gas as ACT's Zero Emissions Target Looms', *RenewEconomy*, 4 March 2020, https://reneweconomy.com.au/evoenergy-turns-to-renewable-gas-as-acts-zero-emissions-target-looms-14788/

11 Finkel, A., 'The Orderly Transition to the Electric Planet',

Australian Government, Australia's Chief Scientist, 12 February 2020, *https://www.chiefscientist.gov.au/news-and-media/national-press-*club-address-orderly-transition-electric-planet

12 *Ibid.*un

13 Powell, D. and Johansson, S., 'Business Council of Australia Hits Back as Pressure Mounts on Climate Change, *Sydney Morning Herald*, 14 January 2020, https://www.smh.com.au/business/companies/business-council-faces-criticism-over-regressive-climate-change-policy-20200114-p53raa.html

14 Garnaut, R., 'Three Policies that Will Set Australia on a Path to 100% Renewables', *Guardian*, 6 November 2019, https://www.theguardian.com/australia-news/2019/nov/06/ross-garnaut-three-policies-will-set-australia-on-a-path-to-100-renewable-energy

15 Osmond, D., 'How to Run the National Electricity Market on 96% Renewables', *RenewEconomy*, 3 March 2020, https://reneweconomy.com.au/how-to-run-the-national-electricity-market-on-96-per-cent-renewables-91522/

16 Casey, J. A. *et al*, 'Coal-fired Power Plant Closures and Retrofits Reduce Asthma Morbidity in the Local Population', *Nature Energy*, 1 May 2020, https://www.nature.com/articles/s41560-020-0622-9

17 Garnaut, R., *Superpower: Australia's Low-Carbon Opportunity*, Latrobe University Press, 2019.

Chapter 8

1 Hendry, M., 'Black Lung Advocates Say 20 Queenslanders Diagnosed with Coal Dust Diseases in a Fortnight' ABC News, 27 February 2019, https://www.abc.net.au/news/2019-02-26/dozens-of-new-black-lung-cases-qld-advocates-say/10851482

2 Appunn, K. *et al*, 'Germany's Energy Consumption and Power Mix in Charts', *Clean Energy Wire*, 31 March 2020, https://www.cleanenergywire.org/factsheets/germanys-energy-consumption-and-power-mix-charts

3 *Ibid.*

4 Campbell, E., 'Germany Is Shutting Down Its Coal Industry for Good, so far without Sacking a Single Worker', ABC News, 18 February, 2020, https://www.abc.net.au/news/2020-02-18/australia-climate-how-germany-is-closing-down-its-coal-industry/11902884

5 Vorrath, S., 'Huge Wind Farm Planned for Victoria's Coal Centre, Overlooking Closed Hazelwood Plant', *RenewEconomy* 28 March, 2019, https://reneweconomy.com.au/huge-wind-farm-planned-for-victorias-coal-centre-overlooking-closed-hazelwood-plant-46221/

6 Campbell, E., 'Germany Is Shutting Down Its Coal Industry

for Good, so far without Sacking a Single Worker', ABC News, 18 February, 2020, https://www.abc.net.au/news/2020-02-18/australia-climate-how-germany-is-closing-down-its-coal-industry/11902884

Chapter 9

1 Morton, A., 'A 60% Rise in Industrial Emissions Points to the Failure of the Coalition's Safeguard Mechanism', *Guardian*, 12 February 2020, https://www.theguardian.com/australia-news/2020/feb/12/a-60-rise-in-industrial-emissions-points-to-failure-of-coalitions-safeguard-mechanism
2 'What the Frack? Australia Overtakes Qatar as World's Largest Gas Exporter', Climate Council, 18 January 2019, https://www.climatecouncil.org.au/australia-worlds-largest-gas-exporter/
3 Australian Government, Department of Environment and Energy, 'Quarterly Update of Australia's Greenhouse Gas Inventory: March 2019', Commonwealth of Australia, 2019, https://www.environment.gov.au/system/files/resources/6686d48f-3f9c-448d-a1b7-7e410fe4f376/files/nggi-quarterly-update-mar-2019.pdf
4 Khadem, N., 'ATO Data Reveals that One Third of Large Companies Pay No Tax', ABC News, 2 January 2020, https://www.abc.net.au/news/2019-12-12/ato-corporate-tax-transparency-data-companies-no-tax-paid/11789048?nw=0
5 Ziffer, D., 'Gas Exports Blamed for Soaring Electricity Prices and Job Losses', ABC News, 17 May 2019, https://www.abc.net.au/news/2019-05-17/gas-exports-blamed-for-electricity-price-rises-job-losses/11121120
6 Long, S., 'Gas Giants Mislead Governments, and It Is Costing Australian Jobs, ACCC Boss Says', ABC News, 27 February 2020, https://www.abc.net.au/news/2020-02-27/gas-giants-misled-governments-accc-boss-rod-sims-says/12004254
7 Parkinson, G., 'Why an Australian Mining Giant Chose Wind and Solar over Gas for $1 Billion Project', *RenewEconomy*, 25 February 2020, https://reneweconomy.com.au/why-an-australian-mining-giant-chose-wind-and-solar-over-gas-for-1-billion-project-15651/
8 'Rio Tinto to Invest $1 Billion to Help Meet New Climate Change Targets', Rio Tinto, 26 February 2020, https://www.riotinto.com/en/news/releases/2020/Rio-Tinto-to-invest-1-billion-to-help-meet-new-climate-change-targets

Chapter 10

1 Vaughan, A., 'Fossil Fuel Divestment: A Brief History', *Guardian*, 9 October 2014, https://www.theguardian.com/environment/2014/oct/08/

fossil-fuel-divestment-a-brief-history

2 Edgecliffe-Johnson, A. and Nauman, B., 'Fossil Fuel Divestment Has "Zero" Climate Impact, Says Bill Gates', *Financial Times*, 17 September 2019, https://www.ft.com/content/21009e1c-d8c9-11e9-8f9b-77216ebe1f17

3 Quiggin, J., 'Adani Beware: Coal Is on the Road to Becoming Completely Uninsurable', *The Conversation*, 13 August, 2019, https://theconversation.com/adani-beware-coal-is-on-the-road-to-becoming-completely-uninsurable-121552

4 Smee, B., 'World's Largest Insurance Broker Under Pressure over Support for Adani and Other Coal Projects', *Guardian*, 3 March 2020, https://www.theguardian.com/business/2020/mar/04/worlds-largest-insurance-broker-under-pressure-over-support-for-adani-and-other-coal-projects

5 Williams, J., 'Carbon Tracker, Coal Developers Risk US$600 Billion', *World Coal*, 12 March 2020, https://www.worldcoal.com/coal/12032020/coal-developers-risk-us600-billion-as-renewables-outcompete-worldwide/

6 Pritchard, M. and Claughton, D., 'Thermal Coal Spot Price Tumbles 25 per cent, Putting Pressure on Some Producers', ABC Rural, 15 May 2020, https://www.abc.net.au/news/rural/2020-05-15/spot-price-for-coal-drops-by-25pc/12241502

Chapter 11

1 Collodi, G., 'Hydrogen Production Via Steam Reforming with CO_2 Capture', *Chemical Engineering Transactions*, 19, 20 April 2010, pp. 37–42, https://www.cetjournal.it/index.php/cet/article/view/CET1019007

2 Finkel, A., 'Orderly Transition to the Electric Planet', Australian Government, Australia's Chief Scientist, 12 February 2020, https://www.chiefscientist.gov.au/news-and-media/national-press-club-address-orderly-transition-electric-planet

3 Sunditch, N., 'Australia Is About to Get Its First Coal-to-Hydrogen Plant in the Latrobe Valley', *Stockhead*, 20 February 2020, https://stockhead.com.au/energy/australia-is-about-to-get-its-first-coal-to-hydrogen-plant-in-the-latrobe-valley/

4 Mazengrab, M., 'Turnbull's Brown Coal Hydrogen Horror Show: $500 million 3 tonnes', *RenewEconomy*, 15 May 2020.

5 Finkel, A., 'The Orderly Transition to the Electric Planet', Australian Government, Australia's Chief Scientist, 12 February 2020, *https://www.chiefscientist.gov.au/news-and-media/national-press-*club-address-orderly-transition-electric-planet

6 Parkinson, G., 'Pilbara Green Hydrogen Project Grows to 15GW Wind and Solar', *RenewEconomy*, 12 July 2019, https://reneweconomy.com.au/

pilbara-green-hydrogen-project-grows-to-15gw-wind-and-solar-97972/
7 Lee, A., 'Australian State Sets 200% Renewable Energy Target to Power Cheap Green Hydrogen', *ReCharge*, 4 March 2020, https://www.rechargenews.com/transition/australian-state-sets-200-renewable-energy-target-to-power-cheap-green-hydrogen/2-1-766988
8 'Jemena's Hydrogen Gas Plant', Smart Energy Council, 20 August 2019, https://www.smartenergy.org.au/news/jemena-s-hydrogen-gas-plant
9 Renewables SA, 'Hydrogen Projects in South Australia', http://www.renewablessa.sa.gov.au/topic/hydrogen/hydrogen-projects
10 'World First in Duisburg as NRW Economics Minister Pinkwart Launches Tests at Thyssenkrupp into Blast Furnace Use of Hydrogen', Thyssenkrupp, 13 November 2019, https://www.thyssenkrupp.com/en/newsroom/press-releases/world-first-in-duisburg-as-nrw-economics-minister-pinkwart-launches-tests-at-thyssenkrupp-into-blast-furnace-use-of-hydrogen-17280.html
11 *Ibid.*
12 Wood, T. *et al*, 'Start with Steel: A Practical Plan to Support Carbon Workers and Cut Emissions', The Grattan Institute, 10 May 2020, https://grattan.edu.au/report/start-with-steel/
13 Delbert, C., 'Hydrogen Powers Commercial Steel Production for the First time', *Popular Mechanics*, 13 May 2020, https://www.popularmechanics.com/science/a32460567/hydrogen-powers-steel-production/
14 Ker, P., 'China's Biggest Steel Maker Explores Hydrogen Substitute', *Australian Financial Review*, 5 March 2020, https://www.afr.com/companies/mining/china-s-biggest-steel-maker-explores-hydrogen-substitute-20200304-p546t7

Chapter 12

1 Newman, P., 'Why Trackless Trams Are Ready to Replace Light Rail', *The Conversation*, 26 September 2018, https://theconversation.com/why-trackless-trams-are-ready-to-replace-light-rail-103690
2 Jolly, J., '2020 Set to be the Year of the Electric Car, Say Industry Analysts', *Guardian*, 25 December 2019, https://www.theguardian.com/environment/2019/dec/25/2020-set-to-be-year-of-the-electric-car-say-industry-analysts
3 Morton, A., 'Electric Car Vehicles Triple in Australia as Sales of Combustion Engine Cars Fall 8%', *Guardian*, 6 February 2020, https://www.theguardi.an.com/environment/2020/feb/06/electric-vehicle-sales-triple-in-australia-as-sales-of-combustion-engine-cars-fall-8. Sam Maclean, Tesla, pers. comm.

4 'Electric Vehicle Outlook 2020', Bloomberg NEF, https://about.bnef.com/electric-vehicle-outlook/
5 Nealer, R. *et al*, 'Cleaner Cars from Cradle to Grave', Union of Concerned Scientists. November 2015, https://www.ucsusa.org/sites/default/files/attach/2015/11/Cleaner-Cars-from-Cradle-to-Grave-full-report.pdf
6 Sam MacLean, Tesla, pers. comm.
7 *Ibid.*
8 Common errors include using an incorrect formula or the wrong Luxury Car Tax (LCT) threshold; dealers/resellers who deferred LCT, not reporting and paying LCT on their BAS immediately after they sell the car or were starting to use it for a non-quotable purpose: primary producers or tourism operators claiming a refund via the BAS and not via the appropriate form: claiming a GST credit for the GST and LCT, when the taxpayer cannot claim back the full GST or the LCT. Sam MacLean, Tesla, pers. comms.
9 Guthrie, S.,'NSW Net Zero Plan: Targeted Electric Car Strategy Revealed', *CarAdvice*, 16 March 2020, https://www.caradvice.com.au/835063/nsw-electric-car-incentives/

Chapter 13

1 Lombrana, L. and Warren, H., 'A pandemic that Cleared Skies and Halted Cities Isn't Slowing Global Warming', *Bloomberg Green*, 8 May 2020, https://www.bloomberg.com/graphics/2020-how-coronavirus-impacts-climate-change/?sref=52ZWO6YM
2 Selkirk, D., 'Is This the Start of the Aviation Revolution?', BBC, 12 February 2020, https://www.bbc.com/future/article/20200211-the-electric-plane-leading-a-revolution
3 *Ibid.*
4 Becken, S. and Pant, P., 'Airline Initiatives to Reduce Climate Impact', Griffith University, 2019, https://www.griffith.edu.au/__data/assets/pdf_file/0028/926506/Airline-initiatives-to-reduce-climate-impact.pdf
5 'How Swedish Airlines Plan to Be Fossil Fuel Free by 2045', *The Local*, 16 April 2019, https://www.thelocal.se/20190416/sweden-airlines-fossil-free-biofuel-climate-compensate-sas
6 *Ibid.*
7 'Proposal for Sweden to Follow Norway's Lead and Mandate Use of Sustainable Aviation Fuels from 2021', *Green Air*, 7 March 2019, https://www.greenaironline.com/news.php?viewStory=2574
8 *Ibid.*
9 Conca, J., 'Carbon Engineering—Taking CO_2 Right Out of the Air to

Make Gasoline', *Forbes*, 8 October 2019, https://www.forbes.com/sites/jamesconca/2019/10/08/carbon-engineering-taking-co2-right-out-of-the-air-to-make-gasoline/#72f7b03513cc
10 PAEHolmes, 'Potential Measures for Air Emissions from NSW Ports', 23 June 2011, https://www.epa.nsw.gov.au/-/media/epa/corporate-site/resources/air/portspreliminarystudy.pdf?la=en&hash=525A0D85D04FD6A192B5544E057A26E5D398D63B
11 California Air Resources Board, 'Ocean-Going Vessels at Berth Regulation', 26 March 2020, https://ww3.arb.ca.gov/ports/shorepower/shorepower.htm
12 Whiting, K., 'An Expert Explains: How the Shipping Industry Can Go Carbon Free', World Economic Forum, 22 September 2019, https://www.weforum.org/agenda/2019/09/an-expert-explains-how-the-shipping-industry-can-go-carbon-free/
13 Radowitz, B., 'World's First Liquid Hydrogen Fuel Cell Cruise Ship Planned for Norway's Fjords', *Recharge*, 3 February 2020, https://www.rechargenews.com/transition/world-s-first-liquid-hydrogen-fuel-cell-cruise-ship-planned-for-norway-s-fjords/2-1-749070
14 Collins, L., 'World's First Liquified Hydrogen Carrier Launched in Japan', *Recharge*, 11 December 2019, https://www.rechargenews.com/transition/worlds-first-liquefied-hydrogen-carrier-launched-in-japan/2-1-722155

Chapter 14

1 Climatenetwork.org/fossil-of-the-day
2 Figueres, C. and Rivett-Carnac, T., *The Future We Choose*, Bonnier, London, 2020.
3 *Ibid.*

Chapter 15

1 O'Donnell, T. and Mummery, J., 'The 2017 Budget Has Axed Research to Help Australia Adapt to Climate Change', *The Conversation*, 11 May 2017, https://theconversation.com/the-2017-budget-has-axed-research-to-help-australia-adapt-to-climate-change-77477
2 'History of Climate Adaptation and Research', CSIRO, https://research.csiro.au/climate/introduction/history-of-climate-adaptation-research-in-csiro/
3 Australian Government, Department of Agriculture, Water and Environment, 'Climate Change', http://www.environment.gov.au/climate-change

4 Barry, M. *et al*, 'Australia Unprepared for Worsening Extreme Weather', Emergency Leaders for Climate Action, Climate Council, 2019, https://www.climatecouncil.org.au/wp-content/uploads/2019/04/fire-chiefs-statement-pages.pdf
5 'Extreme Heat Damaging Health and Livelihoods, and Threatening to Overwhelm World's Hospitals', University of Exeter, 28 November 2018, https://www.exeter.ac.uk/news/research/title_694672_en.html, https://www.lancetcountdown.org/about-us/
6 Somerville, E., 'Australia Unprepared for More Frequent Heatwaves, Health and Emergency Services Authorities Say', ABC News, 5 August 2019, https://www.abc.net.au/news/2019-08-05/australia-unprepared-for-frequent-heatwaves-authorities-say/11382040
7 *Ibid*.
8 'Wrong Climate for Big Dams', International Rivers, https://www.internationalrivers.org/resources/wrong-climate-for-big-dams-fact-sheet-3373
9 Hannam, P. and Bower, M., 'The Early Warning System Being Developed to Shore Up Australia's Beaches', *Sydney Morning Herald*, 15 February 2020, https://www.smh.com.au/environment/weather/the-early-warning-system-being-developed-to-shore-up-australia-s-beaches-20200213-p540jo.html
10 Environmental Defenders Office/Bushfire Survivors for Climate Action, 'Bushfire Survivors Launch Legal Action Against EPA', *MirageNews*, 20 April 2020, https://www.miragenews.com/bushfire-survivors-launch-legal-action-against-epa/
11 Readfearn, G., 'Artificial Fog and Breeding Coral: Study Picks Best Great Barrier Reef Rescue Ideas', *Guardian*, 16 April, 2020, https://www.theguardian.com/environment/2020/apr/16/brightening-clouds-and-coral-larvae-study-picks-best-great-barrier-reef-rescue-ideas
12 Lenton, T. *et al*, 'Climate Tipping Points—Too Risky to Bet Against', *Nature*, 575: 592–5, 9 April 2020, https://www.nature.com/articles/d41586-019-03595-0
13 Beer, C. *et al*, 'Protection of Permafrost Soils from Thawing by Increasing Herbivore Density', *Nature*, Scientific Reports, 10: 4170, 17 March 2020, https://www.nature.com/articles/s41598-020-60938-y. Berardelli, J., 'We Could Release Herds of Animals in the Arctic to Fight Climate Change, Says Study', *Science Alert*, 21 April 2020, https://www.sciencealert.com/releasing-herds-of-animals-in-the-arctic-could-help-fight-climate-change-says-study.
14 Crutzen, P. J., 'Albedo Enhancement by Stratospheric Sulfur Injections: A Contribution to Resolve a Policy Dilemma?', *Climatic Change*, 77, 211, 25 July 2006, https://link.springer.com/content/pdf/10.1007/s10584-006-9101-y.pdf

15 Boettcher, M. and Schafer, S., 'Reflecting upon 10 Years of Geoengineering Research: Introduction to the Crutzen +10 Special Issue', AGU, 18 January 2017, https://agupubs.onlinelibrary.wiley.com/doi/full/10.1002/2016EF000521

16 'Global Effects of Mount Pinatubo', Earth Observatory, NASA, 15 June 2001, https://earthobservatory.nasa.gov/images/1510/global-effects-of-mount-pinatubo'

Chapter 16

1 Allen, M. R. *et al*, 'Global Warming of 1.5°C: An IPCC Special Report on the Impacts of Global Warming of 1.5°C above Pre-Industrial Levels and Related Global Greenhouse Gas Emission Pathways, in the Context of Strengthening the Global Response to the Threat of Climate Change, Sustainable Development, and Efforts to Eradicate Poverty', IPCC, 2018, https://www.ipcc.ch/sr15/download/

2 McLaren, D., 'A Brief History of Climate Targets and Technological Promises', *Carbon Brief*, 13 May 2009, https://www.carbonbrief.org/guest-post-a-brief-history-of-climate-targets-and-technological-promises

3 Flannery, T., *Sunlight and Seaweed*, Text Publishing, Melbourne, 2017.

4 Taylor, L. *et al*, 'Enhanced Weathering Strategies for Stabilizing Climate and Averting Ocean Acidification', *Nature Climate Change*, 6: 402–06, 14 December 2015, https://www.nature.com/articles/nclimate2882?draft=collection

5 'A Global Opportunity for Cities to Lead', C40 Cities, https://www.c40.org/why_cities

6 Rissman, J., 'Concrete Change: Making Cement Carbon-Negative', *GreenBiz*, 6 December 2018, https://www.greenbiz.com/article/concrete-change-making-cement-carbon-negative

7 *Ibid.*

Index